STANDARD METHODS FOR THE ANALYSIS OF OILS, FATS AND DERIVATIVES

INTERNATIONAL UNION OF PURE AND APPLIED CHEMISTRY
APPLIED CHEMISTRY DIVISION
COMMISSION ON OILS, FATS AND DERIVATIVES

STANDARD METHODS FOR THE ANALYSIS OF OILS, FATS AND DERIVATIVES

6th EDITION
PART 1 (SECTIONS I AND II)

Prepared for publication by

C. PAQUOT
Centre National de la Recherche Scientifique, F-94320 Thiais, France

PERGAMON PRESS

OXFORD · NEW YORK · TORONTO · SYDNEY · PARIS · FRANKFURT

U.K.	**Pergamon Press Ltd., Headington Hill Hall, Oxford OX3 0BW, England**
U.S.A.	**Pergamon Press Inc., Maxwell House, Fairview Park, Elmsford, New York 10523, U.S.A.**
CANADA	Pergamon of Canada, Suite 104, 150 Consumers Road, Willowdale, Ontario M2J 1P9, Canada
AUSTRALIA	**Pergamon Press (Aust.) Pty. Ltd., P.O. Box 544, Potts Point, N.S.W. 2011, Australia**
FRANCE	**Pergamon Press SARL, 24 rue des Ecoles, 75240 Paris, Cedex 05, France**
FEDERAL REPUBLIC OF GERMANY	**Pergamon Press GmbH, 6242 Kronberg-Taunus, Pferdstrasse 1, Federal Republic of Germany**

First edition 1979

British Library Cataloguing in Publication Data

International Union of Pure and Applied Chemistry.
Commission on Oils, Fats and Derivatives
Standard methods for the analysis of oils, fats
and derivatives.
Part 1 (Sections 1 and 2). - 6th ed.
1. Oils and fats - Analysis - Standards
I. Title II. Paquot, Charles
664'.3 TP671 78-40305

ISBN 0 08 022379 6

In order to make this volume available as economically and as rapidly as possible the author's typescript has been reproduced in its original form. This method unfortunately has its typographical limitations but it is hoped that they in no way distract the reader.

Printed in Great Britain by A. Wheaton & Co. Ltd., Exeter

COMMISSION ON OILS, FATS AND DERIVATIVES

FORMER PRESIDENTS

1930-1933: S. Fachini (Italy)
1933-1936: F. Rivals (France)
1936-1938: V. Vesely (Czechoslovakia)
1938-1940: H. P. Kaufmann (Germany)
1946-1949: G. L. Voerman (Netherlands)
1949-1951: K. A. Williams (United Kingdom)
1951-1953: F. D. Snell (United States)
1953-1957: J. Vizern (France)
1957-1961: H. K. Sturm (Switzerland)
1961-1964/65: G. Wolff (France)
1964/65-1969: H. A. Boekenoogen (Netherlands)
1969-1973: E. Heinerth (Federal Republic of Germany)
1973-1977: H. J. Vos (Netherlands)
1977- : C. Paquot (France)

COMPOSITION OF THE COMMISSION

1/1/1978

TITULAR MEMBERS

President: C. Paquot (France)
Vice President: E. Delvaux (Belgium)
Secretary: A. Hautfenne (Belgium)
H. Brüschweiler (Switzerland)
D. Firestone (United States)
Ö. Levin (Sweden)
J. Pokorny (Czechoslovakia)
H. Wendt (Federal Republic of Germany)

ASSOCIATE MEMBERS

T. Asahara (Japan)
Mrs J. Beare Rogers (Canada)
C. Carola (Italy)
J. Cornelius (United Kingdom)
J. Martinez Moreno (Spain)
A. Petersen (Denmark)
A. Rutkowski (Poland)
J. van der Well (Netherlands)
H. Wessels (Federal Republic of Germany)
J.-P. Wolff (France)

NATIONAL REPRESENTATIVES

Australia:	A. Johnson
Austria:	G. Zwerenz
Belgium:	Mrs B. Jacobsberg
Bulgaria:	D. Chobanov
Canada:	B. Craig
Denmark:	A. Møller
Federal Republic of Germany:	M. Teupel
France:	M. Naudet
Greece:	V. Kapoulas
Hungary:	Mrs. E. Kurucz
India:	M. Chakrabarty
Ireland:	B. McGwynne
Italy:	R. Monacelli
Japan:	T. Hashimoto
Netherlands:	P. Hendrikse
New Zealand:	F. Shorland
Norway:	T. Grimsvang
Poland:	W. Zwierzykowski
Spain:	J. Gracian Tous
Sweden:	R. Ohlson
Switzerland:	A. Diffenbacher
Turkey:	A. Karabatur
United Kingdom:	K. Williams
United States:	W. Link

OBSERVERS

N. Embree (United States)
U. Engelrud (Denmark)
M. Gassiot Matas (Spain)
J. Holló (Hungary)
Mrs E. Lewkowitsch (United Kingdom)
T. Nagai (Japan)
A. Pattinson (New Zealand)
E. Tiscornia (Italy)
H. Vos (Netherlands)

IUPAC Secretariat: Bank Court Chambers, 2-3 Pound Way,
Cowley Centre, Oxford OX4 3YF, U.K.

Foreword

This publication of the International Union of Pure and Applied Chemistry includes the methods of analysis which have been adopted and edited by the Commission on Oils, Fats and Derivatives in Section I, "Oleaginous Seeds and Fruits", and Section II, "Oils and Fats". The Commission will soon publish the methods comprised in the other Sections, as Part 2 of the publication.

The texts have been revised and brought up to date, in line with ISO Recommendations.

This is the sixth Edition of the methods of the Commission, after five preceding Editions: the first in 1930, the second in 1938, the third in 1948, the fourth in 1954, and the fifth in 1964 (with 4 Supplements). In the first, second and third Editions there were only French texts, but the fourth and fifth Editions included both French and English texts : in cases of dispute, the French text was to be taken as the official text. The present sixth Edition is in English only, but translations are contemplated. A French translation is expected to be published shortly.

Some methods of the fifth Edition were omitted, because they are out of date. A new numbering is adopted; it is more valuable than the fifth Edition numbering, and able to receive new methods at a rational place.

It is necessary to state that all the methods have been studied in collaborative tests, and adopted only when concordant results have been obtained.

The Commission on Oils, Fats and Derivatives of the I.U.P.A.C.

It seems highly appropriate to reproduce here the following account, written in 1959 by J. Vizern and published as a preface to the Fifth Edition of the *Standard Methods for the Analysis of Oils, Fats and Soaps*, when the Commission was known as the Oils and Fats Division.

THE OILS AND FATS DIVISION OF THE I.U.P.A.C.

Founded 29 years ago, since when it has had many names and many forms but has remained unchanged in spirit and purpose, the Oils and Fats Division-now within the framework of the International Union of Pure and Applied Chemistry-has won distinction which will again be enhanced by the publicizing of the methods which it has adopted so far.

Its task consisted essentially in the unification of the analytical methods indispensable to the manufacture of, and trade in, fat products; it has subjected the methods employed to a searching examination based exclusively on the results of analyses of identical samples carried out by its members. Some of these methods could be adopted rapidly, some had to be tested and refined by means of patient research, others are being, or will be, studied further, still others have finally been rejected.

Publications prior to and including that of 1954, comprising the texts issued and accepted over many years, needed re-editing in a more uniform style; further, new methods have been adopted since 1954 so that a new publication is called for. It cannot be a final version, since the work is continuing, but it would render the services which one may rightly expect from standardization.

The first attempt at unification was carried out in 1924. In April of that year Professor Fachini, Director of the Experimental Station for the Oils and Fats Industry, invited a number of specialists to Milan to discuss an analytical paper drawn up by an Italian commission whose work he had directed.

In addition to the Italian chemists, Mr. G. Wolff and I represented France, and Dr. A. Gruen, at that time Director of the Schichtwerke Laboratory, Aussig, represented Czechoslovakia. Germany was not represented, as Professor Helde was taken ill and had to abandon his journey. Great Britain's interest in the Conference was shown only by the submission of a paper from the Norman Tate Laboratory, Liverpool.

No decision could emerge from so small a meeting which, moreover, the participants attended in a private capacity. Nevertheless, in the course of the two-day meeting no fundamental differences had arisen between the points of view and, consequently, the path towards unification was clear.

In fact, no progress was made in the first few years, since the attempt made in Milan was not repeated. And yet it deserved to be, for scientific language should be both unequivocal and international with regard to the definitions it uses as well as the techniques which these imply. Should this double character be lacking, difficulties arise in international contracts which both parties sign in good faith but interpret differently. For example, "density" may be interpreted with respect to water at 15°, or as "specific mass" related to water at 4°.

A contract stipulates the conditions of sale of oil after deduction of impurities: one party will determine the insoluble part in an organic solvent which leaves any soaps present on the filter, whereas the other party will regard only the mineral constituents of these soaps as impurities.

A contract specifies a fat product with an acidity not exceeding a certain limit. One party expresses acidity in terms of oleic acid, the other in terms of lauric acid. At first the product will not meet the specification, even though the other party is satisfied. No matter how improbable such discord may seem, it is not unknown, and I can recall a case of arbitration concerning the acidity of illipé seeds sold by a Dutch firm to a French firm. The arbitrators were unable to agree and the chief arbitrator then had to apply a Solomonian judgement to the effect that the mean molecular weight of the illipé fatty acids could be used to evaluate the acidity of this product in terms of lauric acid.

The same applies to the procedures used, for in every analytical determination carried out on complex organic mixtures there are empirical elements. The slightest modification in procedure will, under identical conditions, give different results, so that it is worth while to work out a standard procedure which, while paying due attention to theory, will give reproducible results.

In spite of these obvious facts, objections to this standardization have been raised by certain people who were under the impression that they had found the only correct method and found the retention of their monopoly profitable. Others again were sceptical. However, in the beginning there were a few people who realized that it would be advantageous to establish agreement not only about definitions but also about results, for analytical differences are harmful to the parties concerned and even more so to their experts.

Standardization does not cause loss of individuality. A procedure, however completely described, does not detract from the skill of those carrying out the determinations, and, moreover, there are numerous special cases in which initiative may play a part. Finally, there is the interpretation of the results and the conclusion, for which no procedure whatever can replace the critical judgement and the experience of the analyst.

Not until November 1929 could discussions on the subject of standardization be renewed. At the Barcelona Congress of the Société de Chimie industrielle française, Professor Veselý of the Czechoslovak Polytechnic at Brno made a plea for standardization and reproached us, and in particular the French members, for obstructing—by our inactivity—the efforts made in Italy and in Central Europe, where committees had been formed.

That decided the matter. Immediately after my return from Barcelona I made an attempt to set up a "French Committee for the Standardization of Analytical Methods for Fat Products and Their Derivatives". Most French chemists who specialized in fat chemistry joined this committee, with Professor Rivals as President and Mr. Halphen as Vice-President. A draft set of methods was submitted to them. In January 1930, after three days of discussions, a compilation of recommended methods for the analysis of fat products and soaps was published.

This new body enabled Professor Fachini, in April 1930, to call a meeting, in Geneva, of the National Committees of the most important European countries and to invite representatives for those countries where National Committees had not yet been established.

Without any official recognition, and supported only by the acknowledged competence of the members representing Germany, France, Italy, The Netherlands, Switzerland and Czechoslovakia, the assembly proclaimed itself the International Commission for the Study of Fat Products. Professor Fachini was appointed President, Professor Steger, of Delft University, Vice-President, and Dr. Sporer, assistant to Professor Fachini, Secretary.

The Commission drew up its rules, and especially those governing its composition—three members at most for each participating country: in addition, technical discussions were immediately started. It is not surprising that at this first meeting, at which the members were not well acquainted with each other and at which everybody tried to overcome the language barrier and the difficulties caused by imperfect translations, only the definition of acidity and the method of its determination were arrived at.

Later it was apparent that the confidence resulting from better mutual understanding improved the output of the meetings and that the slow progress was due less to the diversity of languages than to differences in mentality and different ways of thinking, and also to habits often nearly amounting to prejudices.

However, at the following meetings: Zurich 1931, Prague 1932, Rome 1933 (at which the Presidency was given to the doyen Professor Rivals, the Vice-Presidency to Professor Franck (Germany) and the Secretariat to myself) and Paris 1934, important results were obtained.

But at the meeting in London 1935 and Lucerne 1936, as a result of internal changes, little progress was made. The United Kingdom, which until then had only been represented by observers, joined the International Commission. Naturally, the procedures previously adopted were submitted to the new member for approval. In addition, the German delegation including Professor Franck and Dr. Rietz which, since Geneva, had been sponsored by the "Wizoeff", was presided over in London by Professor Täufel and later, in Lucerne, by Professor Kaufmann, the "Wizoeff" having become the Deutsche Gesellschaft für Fettforschung.

Moreover, apart from the technical discussions, part of the sessions was occupied in working out statutes which had become necessary in view of the possible affiliation to I.U.P.A.C., which was achieved at the Lucerne meeting.

At Lucerne, the Presidency of the Commission was given to Professor Veselý, and Professor Kaufmann was appointed Vice-President. The International Commission submitted to I.U.P.A.C. the internationally adopted specifications relating to the following methods:

Water and Volatile Constituents, Impurities, Ash, Acid Value, Saponification Value, Iodine Value, Density, Refractive Index, Unsaponifiable Matter.

It was not much to show after six years, but this only stresses the difficulties attached to the enterprise. However, with the support of the I.U.P.A.C. and having attained a certain cohesion and a spirit of confidence and even of friendship among its members, which makes for better mutual understanding, the Commission could look forward to greater efficiency in future. A working meeting was held at Paris in 1937, and at Rome in 1938 where Professor Kaufmann was elected President. It was intended to meet again in Berlin in September 1939, but the war occasioned a long and painful break.

The resumption of the meetings took place in London in 1946. The Commission was established again under the Presidency of Dr. G. L. Voerman (Netherlands), whose initiative led to integration of the Commission into I.U.P.A.C. as a sub-section of the Section of Applied Chemistry. The Secretariat was again entrusted to me. At this new start the Commission comprised members from: Denmark, the U.S.A., France, the U.K., Italy, The Netherlands, Switzerland and Czechoslovakia.

The activities of the Commission were resumed at the London meeting in 1947. The meeting envisaged for Paris in 1948 was cancelled. We then thought it would be sufficient to meet every two years, at the same time as the I.U.P.A.C., but since the Amsterdam meeting in 1949 the annual routine has been resumed as it was considered that the efficiency would be adversely affected by long periods without contact. At Amsterdam, Spain was admitted as a new member. Dr. Voerman, the President, having expressed his desire to retire from his office, was replaced by Dr. K. A. Williams, with Mr. Foster Dee Snell (U.S.A.) and Dr. H. Sturm (Switzerland) as Vice-Presidents.

Then came Paris, 1950, and New York, 1951. In order to bring its regulations into line with the statutes of the Union, a Sub-Commission was set up, consisting of not more than ten members. This Sub-Commission represented the Commission which was incorporated in the I.U.P.A.C. Foster Dee Snell became President and H. Sturm Vice-President.

At Oxford, 1952, the Commission changed its name and became the Oils and Fats Division of the Section of Applied Chemistry of the I.U.P.A.C.; and the Sub-Commission became the Division Committee.

Austria, which before the war had been represented by Dr. Halden as observer, and Belgium, which had meanwhile been admitted, became members of the Division at the Stockholm meetings in 1953. The present author was appointed President and Dr. H. Sturm Vice-President, while the Secretariat was entrusted to Mr. J.-E. Bertrand (Belgium). It was important for the Secretary to have a perfect knowledge of French, for from the beginning it had been decided that the authorized versions should be worded in this language. Since that time the correctness of the choice has been demonstrated by his activities and the efficiency with which he has dealt with the increasing amount of work which has resulted from the participation of more and more countries.

Germany, which, in the circumstances, had not taken part since 1939, again joined the Commission at Seville in 1954.

The Division Committee submitted some modifications because of the limitations on the periods of office imposed by the statutes of I.U.P.A.C. Dr. S. H. Bertram (Netherlands) became Vice-President.

Then came the meeting in Zurich in 1955 and in Copenhagen in 1956.

In 1957, at Paris, Dr. H. K. Sturm was elected President, with Professor Jacini (Italy) Vice-President.

In 1958, at Brussels, Ireland was admitted to the Division. The Argentine, which had been represented by Professor Cattaneo for three years, again became a member. Fourteen countries are, therefore, represented. Sweden and Israel are on the waiting list for membership.

In 1959, at Munich, the Division was in a position to submit to the Bureau of I.U.P.A.C., with a view to publication, more than forty test methods relating to oil seeds, fat products, glycerides and soaps, which had been subjected to a very strict selection after long co-operative studies and comparisons. The compendium of these methods will constitute a work of reference. The door has been opened for the admittance of new trends. Thus, after five years of, at times, heated discussions, struggling against ingrained habits which could not apparently be broken, the Division has now rejected subjective colour determinations in favour of the photo-electric eye of the spectrometer. The document is not final; work is still progressing, certain methods can be modified, others are being studied.

No human task is perfect nor ever finished, but all those who have participated in this work may be proud. Towering above them all is Professor Fachini, who initiated the project and who has inspired all. The thirtieth anniversary of the Division will be celebrated in 1960, in his country.

Some members have passed away. My thoughts go out to the memory of Professor Rivais and of Dr. Voerman, both of whom presided so ably.

But a team continues to operate. The names of those who compose it are a guarantee of its quality.

November 1959

J. VIZERN

Former Vice-President of the Applied Chemistry Section of I.U.P.A.C.
Former Chairman of the Oils and Fats Division of the I.U.P.A.C.

Since 1959, nearly 20 years have passed and the administrative history of those years may be briefly summarized.

Annual meetings of the Division have been held: after meeting at Bari (1960), London (1961) and Prague (1962), the Division met again in London (1963) when H.J. Vos replaced J. Bertrand as Secretary, and the "Oils and Fats Division" was renamed the "Oils and Fats Section". Later, two of the founder members retired: G. Wolff in 1964 at Rotterdam and J. Vizern at Paris (1965).

Meetings were held at Dublin (1966), Prague (1967), Vienna (1968) and at Cortina d'Ampezzo in 1969: at this meeting the "Conseillers" became "Titular Members" and "Associate Members" were designated.

Stockholm was the venue in 1970 and St Gallen in 1971, when, at the behest of I.U.P.A.C., the Oils and Fats Section was divided into two Commissions: the "Commission for Oils and Fats" and the "Commission for Soaps and Oleochemicals", the Titular Members being the same in both Commissions.

In 1972 the meeting was held at Chester, and at Munich in 1973, H.J. Vos became President and was succeeded as Secretary by C. Paquot.

Subsequent meetings were held at Jablonna (1974) and Madrid (1975), and the "Section" was re-entitled the "Oils and Fats Commission"; the two former Commissions were reunited in the Commission, which has to conform progressively to the bye-laws of IUPAC: eight Associate Members (but twelve until 1979) and one National Representative per country (two until 1977) are allowed.

Following Lyngby in 1976, the 1977 meeting was held at Warsaw: the President, H.J. Vos, yielded the chair to C. Paquot and A. Hautfenne became the new Secretary. The Commission altered its name to "Commission on Oils, Fats and Derivatives".

Since 1959 the number of adhering countries of the Commission has increased: Hungary and India joined in 1971, as well as Nigeria (until 1976); in 1972 Canada, in 1973 Japan and New Zealand, in 1974 Bulgaria, in 1975 Norway, in 1976 Australia and in 1977 Greece and Turkey joined in the work. Argentina, however, is no longer a member of the Commission.

Over the years, the Commission has been saddened by the deaths of its last founder members: Professor Vessely in 1964, J. Vizern in 1968 and G. Wolff in 1969, but the memory of them will always remain green, because the Commission owes so much to them.

It is useful to record that, since 1959, more than thirty new methods have been added to the fifty or so previously published. The work continues: twelve methods are now under study; likewise, the revision of Sections III (Glycerines) and IV (Soaps) proceeds, while the preparation of further Sections is in hand.

So we have good reason to claim that the Commission on Oils, Fats and Derivatives continues its efforts to be useful to all.

C. PAQUOT

President of the Commission on
Oils, Fats and Derivatives.

Contents

Warning

Some of the methods require the utilization of particularly flammable and/or poisonous solvents.

In this case the safety instructions for the handling of these solvents should be observed. The procedure must be carried out in a well-ventilated room, preferably in a fume cupboard, to prevent the exposure of laboratory workers to poisonous vapours.

Section I
Oleaginous Seeds and Fruits

1.001. Preparation of the Sample for Analysis

(Fifth Edition: Method I.A.1)

1. SCOPE

This Standard describes the procedures for obtaining, from a laboratory sample of oleaginous seeds or fruits, a sample for analysis.

2. FIELD OF APPLICATION

This Standard is applicable to all oleaginous seeds and fruits.

3. DEFINITIONS

Laboratory sample: sample in the state of preparation in which it is received by the laboratory;

Test sample: sample in the state of preparation in which it is analysed;

Test portion: quantity of product withdrawn from the test sample (or, if it is the same, from the laboratory sample) and used in the analysis.

4. PRINCIPLE

Preparation of the test sample from the contract sample (laboratory sample) by suitable means, if necessary after the removal of impurities of large size, using one or other of the dividing instruments specified and taking care that the final test sample truly represents the bulk of the laboratory sample.

5. APPARATUS

5.1. Dividing apparatus: quartering apparatus, conical divider, or other dividing apparatus which will ensure uniform distribution of the components of the laboratory sample in the test sample.

6. DIVISION OF LABORATORY SAMPLE

After separating and weighing the impurities of large size, if necessary, mix the laboratory sample carefully in order to make it as uniform as possible, and, by means of dividing apparatus (5.1) appropriate to the nature of the seed, reduce it successively until there is obtained approximately the mass of material shown in the next table:

Species of seeds or fruits	Minimum mass of each test sample (g)
Copra, olives	1000
Castor, palm kernel, groundnut	600
Shea nut, sunflower, soja, cottonseed	500
Linseed, rape, rubsen, poppy, mustard, sesame	200

Note. For seeds not included in the above table, the minimum quantities should be the same as prescribed for seeds of similar size.

Section I
Oleaginous Seeds and Fruits
Determination of the Physical Characteristics

1.101. Determination of the Impurities

(Fifth Edition: Method I.A.2)

1. SCOPE

This Standard describes a method for the determination of impurities in oleaginous seeds and fruits and defines the various categories of impurities.

2. FIELD OF APPLICATION

This Standard is applicable to all oleaginous seeds and fruits.

3. DEFINITIONS

By *impurities* are meant all foreign matter, organic and inorganic, other than seeds or fruits of the basic species.

The following are considered as impurities:

- *fines*, particles passing the sieves shown in (5.1);
- *large non-oleaginous foreign bodies* (wood debris, stones, pieces of metal) and those which are part of the seed (fragments of free or adhering husks);
- *oleaginous impurities*, i.e. oleaginous seeds other than seeds of the basic species.

4. PRINCIPLE

Separation of the impurities by sieving and sorting into three categories: fines, non-oleaginous impurities, oleaginous impurities. Determination of the mass of each category.

5. APPARATUS

5.1. Sieves, with circular holes, diameter of holes in mm:

Copra	2.0
Small seeds (linseed, rapeseed,...)	0.5
All other oleaginous seeds	1.0

5.2. Tweezers, or other suitable instruments.

6. PROCEDURE

The test portion is the analysis sample obtained according to 1.001.

Weigh the test portion with a precision of at least 0.1 per cent.

The determination of impurities should be carried out sufficiently quickly to avoid any appreciable change in the moisture content.

6.1. Fines

Separate the fines quantitatively by sieving the test portion with the correct sieve (5.1). Collect the fines and weigh them to the nearest 0.01 g.

6.2. Large Non-oleaginous Foreign Bodies

In the material retained by the sieve in 6.1, separate by means of tweezers (5.2) the large non-oleaginous bodies, detaching if necessary bits of shell adhering to the seeds. Weigh them to the nearest 0.01 g.

6.3. Oleaginous Impurities

Separate by means of tweezers (5.2) the oleaginous impurities. Weigh them to the nearest 0.01 g *(Note 1)*.

6.4. Note Concerning Small Seeds

The determination of the impurities in small seeds, if it is required, includes:

(a) on the whole of the test sample:

- the determination of impurities larger than the seeds (possibly large foreign oil seeds); it is made on the total test sample by screening with the aid of a sieve which allows the seeds to pass through;
- the determination of the dust with the aid of a sieve which does not pass the seeds;

(b) on an aliquot of not less than 10 g of the test portion:

- the sorting of the non-oleaginous dust approximately the same size as the seeds, and the sorting of the small foreign seeds.

7. EXPRESSION OF RESULTS

The content, as a percentage by mass, of each category of impurities is given by Table 1.

Table 1

	Mass (g)	Content (per cent)
Test portion	m_0	
Total dust	m_1	$P = \frac{100\, m_1}{m_0}$
Oil content of total dust (determined according to 1.122)		h
Oil content of clean oil seeds (determined according to 1.122)		H
Non-oleaginous dust		$P_S = P(1 - \frac{h}{H})$
Seed dust		$P_G = P - P_S$
Large non-oleaginous foreign bodies	m_2	$I_N = \frac{100\, m_2}{m_0}$
Oleaginous impurities	m_3	$I_O = \frac{100\, m_3}{m_0}$
Total non-oleaginous impurities		$I_S = P_S + I_N$
Total impurities		$I_T = P_S + I_N + I_O$

Show the results to two decimal places for contents not exceeding 0.5 per cent, and to one decimal place for content above this limit.

8. NOTE

1. Some seeds are difficult to separate (e.g. cruciferous oil seeds).

1.111. Grinding

(Fifth Edition: Method I.A.3)

1. SCOPE

This Standard gives some general rules for the grinding of oleaginous seeds and fruits.

2. FIELD OF APPLICATION

This Standard is applicable to all oleaginous seeds and fruits.

3. PRINCIPLE

In order to obtain representative test portions for the following various analyses, rasp or grind the laboratory sample until the production of a fine meal (maximum dimensions of the grains not more than 2 to 3 mm), avoiding reducing the sample to an impalpable powder, generating heat and producing significant loss of moisture.

4. APPARATUS

4.1. Mechanical crushing mill, easy to clean, suitable for the type of seeds to be analysed, and capable of grinding these without heat production or significant loss of moisture and oil.

4.2. Mechanical rasp, or, failing this, hand rasp.

5. PROCEDURE

5.1. For Copra

Rasp the laboratory sample by means of the mechanical rasp (4.2) or, failing this, the hand rasp. If a hand rasp is used, the total amount of the laboratory sample cannot be rasped, and consequently a sub-sample has to be produced, which is as representative as possible. To this effect, the size and the colour of the different particles have to be taken into account. The grains of the rasped material may be longer than 2 mm, but must not exceed 5 mm.

Mix the raspings carefully, and carry out the determinations without delay.

5.2. For Medium-size Seeds (e.g. Sunflower, Peanut, Soyabean)

Grind the test sample in the mechanical crushing mill (4.1) which has previously been cleaned carefully, until the production of particles which are not longer than 2 mm in their longest dimension. Reject the first run of the meal (about one-twentieth of the sample), collect the remainder, mix it carefully and carry out the determinations without delay.

5.3. For Small Seeds (e.g. Linseed, Rapeseed)

For small seeds the determinations are carried out without mechanical grinding. Mix the test sample carefully.

Section I
Oleaginous Seeds and Fruits
Analysis of the Seeds and Fruits

1.121. Determination of Moisture and Volatile Matter Content

(Fifth Edition: Method I.B.1)

1. SCOPE

This Standard describes a method for the determination of moisture and volatile matter content in oleaginous seeds and fruits.

2. FIELD OF APPLICATION

This Standard is applicable to all oleaginous seeds and fruits.

3. DEFINITION

The moisture and volatile matter content of oleaginous seeds and fruits is the loss in mass which the product to be analysed undergoes on heating at 103°C under the operating conditions specified below; it is reported as a percentage of the seeds as received, i.e. containing impurities.

If required, the moisture and volatile matter content in the pure seeds (i.e. those from which the impurities and oily dust have been removed) can be reported. In this case, the certificate of analysis should state "moisture and volatile matter content of pure seeds".

4. PRINCIPLE

Heating the test portion obtained according to 1.111 in an oven at 103 ± 2°C.

5. APPARATUS

5.1. Metal dishes, 7 cm diameter and 3 to 4 cm deep, with lids.

5.2. Electrical oven, regulated at 103 ± 2°C.

5.3. Desiccator, with an efficient dehydrator.

6. PROCEDURE

Weigh in a previously tared dish with its lid, to within 0.001 g, either 5 ± 0.5 g of raspings (1.111.5.1) for copra, or meal (1.111.5.2) for medium-size seeds, or 5 to 10 g of whole seeds for small seeds.

Place the dish, without its lid, into the oven (5.2) and heat at 103 ± 2°C for 3 hours.

Open the oven, and close immediately the dish with its lid. Cool in the desiccator (5.3). Weigh.

Heat in the oven for a further period of 1 hour. Cool. Weigh.

If the difference between the two weighings is less than 5 mg, stop the operation. If not, repeat the heating for successive 1-hour periods until the difference between two successive weighings is less than 5 mg.

Carry out two determinations on the same test sample.

7. EXPRESSION OF RESULTS

The moisture and volatile content, as a percentage by mass, is given by the formula:

$$\frac{100\, m}{m_0}$$

where m is the loss of mass, in g, determined from the final weight,
m_0 is the mass, in g, of the test portion.

Report the result to one decimal place.
Take as the result the arithmetic mean of the two determinations.

1.122. Determination of Oil Content (Extraction Method)

(Fifth Edition: Method I.B.2)

1. SCOPE

This Standard describes an extraction method for the determination of oil content in oleaginous seeds and fruits.

2. FIELD OF APPLICATION

This Standard is applicable to all oleaginous seeds and fruits.

3. DEFINITION

Oil is the whole of the substances extracted under the operating conditions specified below.

The oil content is expressed as a percentage by mass of the seeds as received, i.e. containing impurities.

If required the oil content can be reported as a percentage of pure seeds (i.e. those from which impurities and dust have been removed). In this case, the certificate of analysis should indicate "oil content in pure seeds".

4. PRINCIPLE

Extraction of the oil in a suitable apparatus with an adequate solvent, *n*-hexane, or, failing this, light petroleum.

5. APPARATUS

5.1. Suitable extraction apparatus, with 250-ml flasks, and electric heating bath.

5.2. Oven.

5.3. Extraction thimbles and cotton wool, fat-free.

5.4. Micro-grinder, or, failing this, mortar and pestle.

5.5. Desiccator.

6. REAGENTS

6.1. *n*-Hexane, or, failing this, light petroleum, b.p. 50-70°C, having a bromine value less than 1. For either solvent, the residue on complete evaporation should not exceed 0.002 g/100 ml.

6.2. Pumice grains, 3 to 4 mm in size, previously dried.

6.3. Sand, previously washed with hydrochloric acid and ignited, if the micro-grinder is not used.

7. PROCEDURE

7.1. Preparation of the Test Portion

The test portion is taken from the test sample prepared according to 1.111.

Weigh, to within 0.01 g, about 10 g of:

- raspings (1.111.5.1) in the case of copra,
- meal (1.111.5.2) in the case of medium-size seeds,
- test sample in the case of small seeds.

In the case of small seeds grind the test portion with the aid of the micro-grinder (5.4), taking care not to leave any seeds intact. Transfer the ground seeds, without loss, into the thimble (5.3), using a spatula. Wipe the bowl of the micro-grinder and the spatula, or the mortar and pestle, with a wad of cotton wool (5.3) soaked with solvent (6.1). Place this wad into the thimble.

If the seeds are very moist (moisture and volatile matter content above 10 per cent, according to 1.121) leave the filled thimble for some time in the oven (5.2) at a temperature not higher than 80°C, to reduce the moisture content to less than 10 per cent.

7.2. Determination

Weigh, to the nearest 0.001 g, two previously dried flasks (5.1) A and B containing 2 or 3 grains of pumice (6.2).

Place the thimble containing the prepared test portion in the extraction apparatus (5.1).

Pour about 150 ml of the solvent (6.1) into flask A. Fit the flask to the extraction apparatus and carry out the heating at such a rate that the solvent in the flask boils gently.

After extracting for 4 hours, allow to cool. Remove the thimble from the extraction apparatus, place it in a current of air in order to evaporate the greater part of the solvent impregnating it.

Empty the thimble into the micro-grinder (5.4) and grind without sand. If the micro-grinder is not used, empty the thimble into the mortar, add about 10 g of sand (6.3) and grind as finely as possible. Replace the mixture into the thimble, and put the latter into the extraction apparatus; replace flask A and extract for 2 hours (second extraction).

Allow to cool, remove the thimble, evaporate the solvent and repeat the grinding as above (without a new addition of sand if a mortar is used). Take flask B and carry out a third extraction for 2 hours with 150 ml of solvent.

Remove the greater part of the solvent from flasks A and B by distillation on a boiling water bath (*note 1*).

Eliminate the last traces of solvent by heating the flasks for about 20 minutes in the oven (5.2) at 103 ± 2°C.

Allow the flasks to cool in the desiccator (5.5) for at least 1 hour, and weigh to the nearest 0.001 g.

Heat again for 10 minutes under the same conditions, allow to cool and weigh.

The difference between these two weighings should not exceed 0.010 g.

If this is not the case, heat again for 10 minutes until the difference in weight is not more than 0.010 g. Record the last weighing of flask A.

If the weight of the oil in flask B is not more than 0.010 g, the operation is complete. If not, using flask B, grind the meal once more and extract again for 2 hours, continuing in the same way until the weight of the oil from the last extraction is not more than 0.010 g. Record the last weight of flask B.

The extracted oil must be clear. If not, determine the impurities with the aid of the solvent (6.1) and deduce the corresponding weight.

8. EXPRESSION OF RESULTS

The oil content, expressed as a percentage by mass of the product as received (*note 2*), is given by the formula:

$$\frac{100\,m}{m_0}$$

where m is the sum of the masses, in g, of oil from the flasks A and B at the last weighing,
m_0 is the mass, in g, of the test portion.

Take as the result the arithmetic mean of two determinations.

Express the result to one decimal place.

9. NOTES

1. In the case of semi-drying or drying oils, it is advisable to remove the residual solvent by drying under reduced pressure.
2. If necessary, the oil content in the dust is determined in the same manner on the whole quantity of the separated dust if its total weight is less than 10 g, or otherwise on a test portion of 10 g.

3. If the foreign oil seeds present have been analysed separately, take the corresponding results into account in the final calculation.

Section I

Oleaginous Seeds and Fruits

Analysis of the Oil Extracted from the Seeds and Fruits

1.151. Determination of Acidity of Oil

(Fifth Edition: Method I.C.1)

1. SCOPE

This Standard describes a method for the determination of the acidity of oil present in oleaginous seeds and fruits.

2. FIELD OF APPLICATION

This Standard is applicable to all oleaginous seeds and fruits.

3. DEFINITION

Acid value and acidity: see 2.201.

4. PRINCIPLE

Determination according to 2.201 of the acid value or the acidity of the oil extracted from seeds as received, or, after separation of dust, of the oil extracted from pure seeds and of the oil of dust, according to 1.122.

5. PROCEDURE

Grind the seeds or fruits according to 1.111, extract as soon as possible oil according to 1.122, and determine as soon as the last weighing of the extraction flasks the acid value or the acidity of the extracted oil according to 2.201.

6. EXPRESSION OF RESULTS

Give the results in acid value or in acidity according to 2.201.

Section II
Oils and Fats

2.001. Preparation of the Sample

(Fifth Edition: Method II.A.1)

1. SCOPE

This Standard describes the treatment which the laboratory sample should undergo before analysis.

2. FIELD OF APPLICATION

This Standard is applicable to animal and vegetable oils and fats.

3. DEFINITIONS

Laboratory sample: sample in the state of preparation in which it is received by the laboratory.

Test sample: sample in the state of preparation in which it is analysed.

Test portion: quantity of product withdrawn from the test sample (or, if it is the same, from the laboratory sample) and used in the analysis.

4. PRINCIPLE

According to the nature of the determinations to be carried out, treatment of the laboratory sample so as to make it homogeneous, to separate substances insoluble in the particular oil or fat, or to eliminate water. The means used for this purpose differ according to the physical state of the product to be analysed.

5. APPARATUS

Electric oven with temperature regulation.

6. REAGENT

Sodium sulphate, anhydrous.

7. PROCEDURE

7.1. Liquid Laboratory Sample, Clear and without Sediment

Render the laboratory sample as homogeneous as possible by turning the container upside down several times.

For some determinations, more particularly determinations in which the results might be affected by the possible presence of water (e.g. determination of the iodine value), the test sample should be previously dried. For this purpose, add anhydrous sodium sulphate in the proportion of 1 to 2 g per 10 g of oil or fat, and keep it in the oven at 50°C. Stir vigorously and filter. The filtrate constitutes the test sample.

7.2. Liquid Sample, Turbid or with Sediment

For some determinations, in particular for the determination of moisture and volatile matter, and impurities insoluble in hexane, turn over the container holding the laboratory sample repeatedly until the sediment is completely separated from the walls of the container and uniformly distributed in the mass of the oil. The test portion should be withdrawn immediately.

For all other determinations, put the container and laboratory sample in the oven at 50°C, keep it until the sample has reached this temperature, and then proceeed as in 7.1. If, after heating and stirring, the sample is not completely clear, filter the oil through a filter paper while in the oven maintained at 50°C. The filtrate should be clear; it constitutes the test sample.

7.3. Solid Laboratory Sample

Melt the laboratory sample by keeping it in the oven, heated to 10°C above the melting point of the fat. If, after the heating, the sample is clear, proceed as in 7.1; if it is cloudy or if it contains a sediment, proceed as in 7.2, the oven being maintained at the same temperature.

Section II

Oils and Fats

Determination of the Physical Characteristics

2.101. Determination of the Density

(Fifth Edition: Method II.B.1)

1. SCOPE

This Standard describes a method for the determination of the density of oils and fats in the liquid state, either directly for oils, or after melting for fats.

2. FIELD OF APPLICATION

This Standard is applicable to liquid oils which do not deposit any solid fat at the temperature of determination.

When the fat is not liquid at 20°C, the density is given at a temperature of 40°C, or 60°C, or if necessary at a higher temperature.

3. DEFINITION

The density of a substance is the ratio of its mass to its volume.
The density of a substance varies with temperature.
For oils and fats it is expressed in grams per millilitre.

4. PRINCIPLE

The mass of a given volume of the oil at the required temperature is determined.

5. APPARATUS

5.1. Pyknometer, preferably of capacity about 50 ml, fitted with a thermometer graduated in 0.1°C, and calibrated at the temperature of measurement (*note 1, note 2*).

5.2. Oven, or thermostat.

6. PREPARATION OF THE SAMPLE

Prepare the sample according to 2.001, "Preparation of the sample".

7. PROCEDURE

Carry out all weighings to the nearest 0.1 mg.

7.1. Procedure for Liquid Fats

Tare the empty pyknometer (5.1) and record its mass m_0.

Fill the dry pyknometer with the prepared sample. When the filled vessel has reached the temperature of the ambient temperature at which the measurement is to be made, a temperature which must be within 5°C of the prescribed temperature (usually 20°C), adjust the volume of the oil to the fixed mark.

Record the temperature of the determination t_1 to the nearest 0.1°C.

Weigh and record the total mass m_1.

7.2. Procedure for Solid Fats

Tare the empty pyknometer (5.1) and record the mass m_0.

Fill the dry pyknometer with the prepared sample, completely melted at a temperature about 10°C above its melting point.

Keep the pyknometer for an hour in the oven or thermostat (5.2) maintained at a temperature which does not differ by more than 5°C from the chosen temperature t; this, moreover, allows any traces of air trapped in the sample to escape.

Adjust the level of the melted fat to the mark.

Record the temperature of the determination t_1 to the nearest 0.1°C.

Allow the pyknometer to cool and wipe it carefully. Weigh the full pyknometer and record the total weight m_1.

8. EXPRESSION OF RESULTS

8.1. Method of Calculation

The density (ρ_t) is given by the formula:

$$\rho_t = \frac{m_1 - m_o}{V_t}$$

where m_o is the mass, in g, of the empty pyknometer,
m_1 is the mass, in g, of the pyknometer filled with fat,
V_t is the volume, in ml, of the fat in the pyknometer at temperature t (*note 2*).

Record as the result the arithmetic mean of at least two determinations expressed to four decimal places.

8.2. Temperature Correction

If the volume of the fat is measured at temperature t_1, near t, correct the result to temperature t by means of the following formulae:

$$\rho_t = \rho_{t_1} + (t_1 - t) \times 0.00068 \quad \text{if} \quad t_1 > t,$$

$$\rho_t = \rho_{t_1} - (t - t_1) \times 0.00068 \quad \text{if} \quad t_1 < t.$$

The correction coefficient 0.00068 is an approximative coefficient which is generally adequate; but if the true coefficient appropriate to the fat under examination is known, it is recommended to use it in order to obtain a more accurate value.

9. REPEATABILITY

The difference between the results of two determinations carried out simultaneously or in a rapid succession by the same analyst shall not exceed 2 units of the fourth decimal place.

10. NOTES

1. A density bottle may be used instead of a pyknometer. If a density bottle is used, when it is filled, take care to insert the stopper in such a way that the capillary portion is completely filled with the fat or water.

2. Calibration of the pyknometer

Tare the empty pyknometer and record its mass m_o.

Fill the pyknometer with water which has been recently distilled and freed from air.

Keep the pyknometer for 1 hour in the oven or thermostat maintained at a temperature within 5°C of the temperature at which the determination should be made (normally 20°C, 40°C or 60°C).

Adjust the water level to the mark.

Record the temperature t_2 of the determination, to the nearest 0.1°C.

If necessary allow the pyknometer to cool to room temperature, carefully wipe the outside dry. Weigh and record its total mass m_2.

The volume V_t of the pyknometer is given by:

$$V_t = \frac{m_2 - m_o}{\rho_{H_2O,t_2}} \left[1 + 0.000025\ (t - t_2)\right]$$

where m_o is the mass, in g, of the empty pyknometer,
m_2 is the mass, in g, of the pyknometer filled with water,
ρ_{H_2O,t_2} is the density of water at t_{20C} (Table 1),
0.000025 is the mean coefficient of cubic expansion of the glass of the pyknometer.

Table 1

Density of water at t_2°C

t_2 in °C	ρ_{H_2O,t_2}	t_2 in °C	ρ_{H_2O,t_2}	t_2 in °C	ρ_{H_2O,t_2}
15	0.99913	32	0.99505	49	0.98852
16	0.99897	33	0.99473	50	0.98807
17	0.99880	34	0.99440	51	0.98762
18	0.99862	35	0.99406	52	0.98715
19	0.99843	36	0.99371	53	0.98669
20	0.99823	37	0.99336	54	0.98621
21	0.99802	38	0.99299	55	0.98573
22	0.99780	39	0.99262	56	0.98525
23	0.99757	40	0.99225	57	0.98475
24	0.99733	41	0.99186	58	0.98425
25	0.99707	42	0.99147	59	0.98375
26	0.99681	43	0.99107	60	0.98324
27	0.99654	44	0.99066	61	0.98272
28	0.99626	45	0.99024	62	0.98220
29	0.99597	46	0.98982	63	0.98167
30	0.99568	47	0.98940	64	0.98113
31	0.99537	48	0.98896	65	0.98059

2.102. Determination of the Refractive Index

(Fifth Edition: Method II.B.2)

1. SCOPE

This Standard describes a method for the determination of the refractive index of animal and vegetable oils and fats.

2. FIELD OF APPLICATION

This Standard is applicable to the animal and vegetable oils and fats.

3. DEFINITION

The refractive index of a medium is the ratio of the speed of light at a definite wavelength *in vacuo* to its speed in the medium.

The refractive index of a given substance varies with the wavelength of the light and with the temperature.

Generally the speed of light in air is taken in the place of the speed *in vacuo*, and, unless otherwise specified, the selected wavelength is the mean wavelength of the D-lines of sodium (589.6 nm). In this case the notation is n_D^t at t°C.

In general the refractive index is given at:

- 20°C for fats which are liquid at this temperature,
- 40°C or 60°C or 80°C, or above, for fats which are solid at a temperature of 20°C.

4. PRINCIPLE

Measurement by means of a suitable refractometer of the refractive index of the liquid sample.

5. APPARATUS

5.1. Refractometer, e.g. Abbe type, able to measure the refractive index to ± 0.0001 within the range n_D = 1.3000 to 1.7000.

5.2. Light source: sodium vapour lamp. White light can also be used if the refractometer is fitted with an achromatic compensation device.

5.3. Glass plate, of known refractive index.

5.4. 100-ml beaker.

5.5. Glass rod, with rounded ends.

5.6. Fluted filter paper.

5.7. Water bath.

5.8. Water bath, thermostatically controlled with circulation pump to maintain the temperature to ± 0.5°C.

6. REAGENTS

6.1. α-Bromonaphthalene.

6.2. Trichlorethylene.

6.3. Sodium sulphate, anhydrous.

7. PREPARATION OF THE SAMPLE

If the sample is liquid at room temperature, mix it thoroughly with a glass rod (5.5), dry it over anhydrous sodium sulphate (6.3) and filter through a fluted filter paper (5.6).

If the fat sample is solid, heat it in a beaker (5.4) on a water bath (5.7) to a temperature 10°C higher than the melting point, dry over anhydrous sodium sulphate, mix thoroughly and filter.

8. PROCEDURE

8.1. Calibration of the Instrument

Verify the regulation of the apparatus (5.1) with a determination of the refractive index of a glass plate (5.3), an optical contact with the prism being made by a drop of α-bromonaphthalene (6.1). If necessary, the system of regulation may be used.

8.2. Measurement

Using a water circulation by the water bath (5.8) and its pump, remain to the nearest 0.5°C the temperature of prisms of the apparatus (5.1) to:

- 20°C for a lecture concerning fats which are fully liquid at this temperature,
- 40°C for a lecture concerning fats which are fully melt at this temperature,
- 60°C for a lecture concerning hydrogenated fats,
- 80°C or above for a lecture concerning waxes.

Keep the sample to the temperature which the measure should be made by immersion of its container into a water bath maintained at this temperature.

Make the determination according to the operational conditions for the used apparatus. Read the refractive index at the nearest 0.0001 in absolute value, and the temperature of the prisms of the apparatus.

Make two other determinations, and take as the result the arithmetic mean of the three readings, rounding to the fourth decimal place if the requirement of repeatability is satisfied.

9. EXPRESSION OF RESULTS

If the difference between the reading temperature t_1 and the prescribed temperature t is lower than 3°C, the refractive index n_D^t for the prescribed temperature t is given by the formula:

$$n_D^t = n_D^{t_1} + (t_1 - t)\,F \quad \text{if } t_1 > t,$$

$$n_D^t = n_D^{t_1} - (t - t_1)\,F \quad \text{if } t_1 < t,$$

where t_1 is the reading temperature,
t is the prescribed temperature,
F is equal to 0.00035 at t = 20°C, or
F is equal to 0.00036 at t = 40°C and t = 60°C.

10. REPEATABILITY

The difference between the results of three determinations carried out in rapid succession by the same analyst shall not exceed 0.0002.

2.103. Spectrophotometric Determination of the Colour

(Fifth Edition: Method II.B.4)

FOREWORD

The subjective methods used for the estimation of the colour of oils and fats are frequently empirical and not easy to carry out; these have not been retained. A more accurate and easier method is required, viz. the determination of the transmission curve between 400 and 700 nm, or measurement of the transmission at definite wavelengths.

Crude oils and fats have transmission curves which are characteristic for each family of fats. All bleached oils and fats, on the contrary, show transmission curves which are similar and without characteristic peaks.

1. SCOPE

This Standard describes a spectrophotometric method for the determinations of the colour of oils and fats by measuring their transmission in the visible spectrum (between 400 and 700 nm).

2. FIELD OF APPLICATION

This Standard is applicable to animal and vegetable oils and fats.

3. PRINCIPLE

Using a spectrophotometer record the transmittance measurements between 400 and 700 nm at regular intervals of 10 nm in cells of a length such that the transmission factor relative to the reference solvent lies between 20 and 80 per cent; do not, however, use cells of less than 0.5 cm or more than 5 cm length, except for special investigations (*note 1*). Determine the position and the extent of the principal inflexions in the curve (the maxima and minima) by measurements made at 1-nm intervals. If necessary, convert the observations to a standard length of 1 cm by calculation.

4. APPARATUS

4.1. 100-ml volumetric flasks.

4.2. Spectrophotometer, provided with cells having parallel plane surfaces 0.5, 1, 2 and 5 (or 4) cm apart.

5. REAGENTS

5.1. Nickel sulphate, analytical reagent quality ($NiSO_4$, $7H_2O$) (*note 2*).

5.2. Cobalt sulphate, analytical reagent quality ($CoSO_4$, $7H_2O$) (*note 2*).

5.3. Carbon tetrachloride, anhydrous, for spectrophotometry.

5.4. Hydrochloric acid, aqueous solution 1 N.

5.5. Reference solution: put 11.5 ml of the aqueous solution of hydrochloric acid (5.4) in the volumetric flask (4.1); dissolve 4.7 g of nickel sulphate (5.1) (Ni = 982 mg) and 3.1 g of cobalt sulphate (5.2) (Co = 650 mg) and make up to 100 ml with distilled water.

6. PREPARATION OF THE SAMPLE

The samples to be examined must be clear. Filter through paper if necessary in the presence of a small quantity of a filter aid which does not affect the colour.

Solid fats should be previously melted and maintained at 10°C above their melting point long enough to ensure their clarity. They must remain clear for the duration of the experimental work. Finally, use a thermostat regulated to the minimum temperature compatible with the melting point of the fat.

Clarify the sample, if necessary, by drying *in vacuo* away from light.

7. PROCEDURE

7.1. Checking the Spectrophotometer

(a) Regulate the wavelength selection device of the spectrophotometer using the following two mercury lines for calibration:

$$\lambda_1 = 546.1 \pm 0.2 \text{ nm},$$

$$\lambda_2 = 435.8 \pm 0.4 \text{ nm}.$$

(b) Check the transmission scale by measuring the transmittance in 1-cm cells using a minimum slitwidth for the observations of the reference solution (5.5) relative to the solvent (5.3). If the apparatus is properly adjusted the following values should be obtained:

$$\lambda = 439 \text{ nm} \qquad T = 51.4 \pm 0.3,$$

$$\lambda = 512 \text{ nm} \qquad T = 26.9 \pm 0.5,$$

$$\lambda = 581 \text{ nm} \qquad T = 73.3 \pm 0.3.$$

(c) Two cells of the same length used alternately for the reference solvent and the reference solution should give, at 550 nm for example, transmittances not differing by more than 0.5 per cent.

7.2. Measurement

Determine the percentage transmission of the sample in a 1-cm cell every 10 nm from 400 to 700 nm compared with that of carbon tetrachloride (5.3) as a blank and using a minimum slitwidth for each observation. At the points of maximum and minimum transmission take readings at 1 nm distance on each side as far as is necessary to secure precision. Make three observations at each wavelength and record the mean. There are two possibilities:

(1) all the readings fall between 20 and 80 per cent, report the readings;
(2) some of the readings fall below 20 or above 80 per cent. For readings falling below 20 per cent, repeat the observations at these wavelengths using a 0.5-cm cell. For readings falling above 80 per cent, but below 90 per cent, repeat the observations using a 2-cm cell. For readings falling above 90 per cent, repeat the observations using a 5- (or 4-) cm cell.

8. EXPRESSION OF RESULTS

By the application of Lambert's law, calculate the value of the transmittance for 1 cm length using the formula:

$$\log \frac{100}{T_1} = \frac{1}{n} \log \frac{100}{T_n}$$

where T_1 is the transmission calculated for a cell of 1 cm,
T_n is the transmission measured for a cell of n cm.

The corresponding values of T and log $100/T$ are given in the tables. Express the results thus obtained either by means of a series of transmittance values at defined wavelengths or by means of a graph plotting the transmission against the wavelength (scale: 1 cm on the abscissa = 20 nm, 1 cm on the ordinate = 5 per cent transmission).

9. NOTES

1. In the case of special investigations, the transmission measurements below 20 per cent can be made in a 1-cm cell on a 1 in 2 volume dilution if the observed transmission is greater than 5 per cent, or in a 1 in 5 volume dilution if it is less.

The dilutions are made by weighing 9.120 g of fat into a standard 20- or 50-ml flask, depending on whether the dilution is to be 1 in 2 or 1 in 5, and then making up to volume with carbon tetrachloride (5.3) at 20°C.

The transmission of a pure fat examined in a 1-cm cell is given by the formula:

$$\log \frac{100}{T_1} = \frac{1}{d} \log \frac{100}{T_d}$$

where d is 1/2 or 1/5,
T_d is the transmission measured at the dilution d.

2. The exact titration of nickel and cobalt in commercial salts may be necessary.

Tables of corresponding value

$T \to \log \frac{100}{T}$

T	0.0	0.1	0.2	0.3	0.4	0.5	0.6	0.7	0.8	0.9
		3.000	2.699	2.523	2.398	2.301	2.222	2.155	2.097	2.046
1	2.000	1.958	1.921	1.886	1.854	1.824	1.796	1.769	1.745	1.721
2	1.699	1.678	1.657	1.638	1.620	1.602	1.585	1.569	1.553	1.538
3	1.523	1.509	1.495	1.481	1.468	1.456	1.444	1.432	1.420	1.409
4	1.398	1.387	1.376	1.365	1.357	1.347	1.337	1.328	1.319	1.309
5	1.301	1.292	1.284	1.276	1.268	1.260	1.252	1.244	1.237	1.229
6	1.222	1.215	1.208	1.201	1.194	1.187	1.180	1.174	1.167	1.161
7	1.155	1.149	1.143	1.137	1.131	1.125	1.119	1.113	1.108	1.102
8	1.097	1.092	1.086	1.081	1.076	1.071	1.065	1.060	1.055	1.051
9	1.046	1.041	1.036	1.032	1.027	1.022	1.018	1.013	1.009	1.004
10	1.000	0.996	0.991	0.987	0.983	0.979	0.975	0.971	0.967	0.963
11	0.959	0.955	0.951	0.947	0.943	0.939	0.936	0.932	0.928	0.924
12	0.921	0.917	0.914	0.910	0.907	0.903	0.900	0.896	0.893	0.889
13	0.886	0.883	0.879	0.876	0.873	0.870	0.866	0.863	0.860	0.857
14	0.854	0.851	0.848	0.845	0.842	0.839	0.836	0.833	0.830	0.827
15	0.824	0.821	0.818	0.815	0.812	0.810	0.807	0.804	0.801	0.799
16	0.796	0.793	0.790	0.788	0.785	0.782	0.780	0.777	0.775	0.772
17	0.769	0.767	0.764	0.762	0.759	0.757	0.754	0.752	0.749	0.747
18	0.745	0.742	0.740	0.738	0.735	0.733	0.730	0.728	0.726	0.724
19	0.721	0.719	0.717	0.714	0.712	0.710	0.708	0.706	0.703	0.701
	0.0	0.1	0.2	0.3	0.4	0.5	0.6	0.7	0.8	0.9

	0.0	0.1	0.2	0.3	0.4	0.5	0.6	0.7	0.8	0.9
20	0.699	0.697	0.695	0.692	0.690	0.688	0.686	0.684	0.682	0.680
21	0.678	0.676	0.674	0.672	0.670	0.668	0.666	0.664	0.662	0.660
22	0.658	0.656	0.654	0.652	0.650	0.648	0.646	0.644	0.642	0.640
23	0.638	0.636	0.634	0.633	0.631	0.629	0.627	0.625	0.623	0.622
24	0.620	0.618	0.616	0.614	0.613	0.611	0.609	0.607	0.606	0.604
25	0.602	0.600	0.599	0.597	0.595	0.593	0.592	0.590	0.588	0.587
26	0.585	0.583	0.582	0.580	0.578	0.577	0.575	0.573	0.572	0.570
27	0.569	0.567	0.565	0.564	0.562	0.561	0.560	0.558	0.556	0.554
28	0.553	0.551	0.550	0.548	0.547	0.545	0.544	0.542	0.541	0.539
29	0.538	0.536	0.535	0.533	0.532	0.530	0.529	0.527	0.526	0.524
30	0.523	0.521	0.520	0.519	0.517	0.516	0.514	0.513	0.511	0.510
31	0.509	0.507	0.506	0.504	0.503	0.502	0.500	0.499	0.497	0.496
32	0.495	0.493	0.492	0.491	0.489	0.488	0.487	0.485	0.484	0.483
33	0.481	0.480	0.479	0.478	0.476	0.475	0.474	0.472	0.471	0.470
34	0.469	0.467	0.466	0.465	0.463	0.462	0.461	0.460	0.458	0.457
35	0.456	0.455	0.454	0.452	0.451	0.450	0.448	0.447	0.446	0.445
36	0.444	0.442	0.441	0.440	0.439	0.438	0.436	0.435	0.434	0.433
37	0.432	0.431	0.430	0.428	0.427	0.426	0.425	0.424	0.422	0.421
38	0.420	0.419	0.418	0.417	0.416	0.414	0.413	0.412	0.411	0.410
39	0.409	0.408	0.407	0.406	0.404	0.403	0.402	0.401	0.400	0.399
	0.0	0.1	0.2	0.3	0.4	0.5	0.6	0.7	0.8	0.9

	0.0	0.1	0.2	0.3	0.4	0.5	0.6	0.7	0.8	0.9
40	0.398	0.397	0.396	0.395	0.394	0.393	0.391	0.390	0.389	0.388
41	0.387	0.386	0.385	0.384	0.383	0.382	0.381	0.380	0.379	0.378
42	0.377	0.376	0.375	0.374	0.373	0.372	0.371	0.370	0.369	0.368
43	0.367	0.366	0.365	0.364	0.363	0.362	0.361	0.360	0.359	0.358
44	0.357	0.356	0.355	0.354	0.353	0.352	0.351	0.350	0.349	0.348
45	0.347	0.346	0.345	0.344	0.343	0.342	0.341	0.340	0.339	0.338
46	0.337	0.336	0.335	0.334	0.333	0.332	0.331	0.330	0.330	0.329
47	0.328	0.327	0.326	0.325	0.324	0.323	0.322	0.321	0.320	0.320
48	0.319	0.318	0.317	0.316	0.315	0.314	0.313	0.312	0.311	0.310
49	0.310	0.309	0.308	0.307	0.306	0.305	0.304	0.303	0.303	0.302
50	0.301	0.300	0.299	0.298	0.297	0.297	0.296	0.295	0.294	0.293
51	0.292	0.291	0.290	0.290	0.289	0.288	0.287	0.286	0.285	0.285
52	0.284	0.283	0.282	0.281	0.280	0.280	0.279	0.278	0.277	0.277
53	0.276	0.275	0.274	0.273	0.272	0.271	0.271	0.270	0.269	0.268
54	0.267	0.267	0.266	0.265	0.264	0.263	0.262	0.262	0.261	0.260
55	0.260	0.259	0.258	0.257	0.256	0.255	0.255	0.254	0.253	0.252
56	0.252	0.251	0.250	0.249	0.249	0.248	0.247	0.246	0.246	0.245
57	0.244	0.243	0.242	0.242	0.241	0.240	0.239	0.239	0.238	0.237
58	0.236	0.236	0.235	0.234	0.233	0.233	0.232	0.231	0.230	0.230
59	0.229	0.228	0.228	0.227	0.226	0.225	0.224	0.224	0.223	0.222
	0.0	0.1	0.2	0.3	0.4	0.5	0.6	0.7	0.8	0.9

	0.0	0.1	0.2	0.3	0.4	0.5	0.6	0.7	0.8	0.9
60	0.222	0.221	0.220	0.219	0.219	0.218	0.217	0.216	0.216	0.215
61	0.214	0.214	0.213	0.212	0.211	0.211	0.210	0.209	0.209	0.208
62	0.207	0.207	0.206	0.205	0.205	0.204	0.203	0.202	0.202	0.201
63	0.201	0.200	0.199	0.198	0.198	0.197	0.196	0.195	0.195	0.194
64	0.194	0.193	0.192	0.192	0.191	0.190	0.189	0.189	0.188	0.187
65	0.187	0.186	0.185	0.185	0.184	0.183	0.183	0.182	0.181	0.181
66	0.180	0.179	0.179	0.178	0.177	0.177	0.176	0.176	0.175	0.174
67	0.174	0.173	0.172	0.172	0.171	0.170	0.170	0.169	0.168	0.168
68	0.167	0.167	0.166	0.165	0.164	0.164	0.163	0.163	0.162	0.162
69	0.161	0.160	0.160	0.159	0.158	0.158	0.157	0.156	0.156	0.155
70	0.155	0.154	0.154	0.153	0.152	0.152	0.151	0.150	0.150	0.149
71	0.149	0.148	0.147	0.147	0.146	0.145	0.145	0.144	0.144	0.143
72	0.143	0.142	0.141	0.141	0.140	0.140	0.139	0.138	0.138	0.137
73	0.136	0.136	0.135	0.135	0.134	0.134	0.133	0.132	0.132	0.131
74	0.131	0.130	0.129	0.129	0.128	0.128	0.127	0.126	0.126	0.125
75	0.125	0.124	0.124	0.123	0.123	0.122	0.121	0.121	0.120	0.120
76	0.119	0.119	0.118	0.117	0.117	0.116	0.116	0.115	0.114	0.114
77	0.113	0.113	0.112	0.112	0.111	0.111	0.110	0.109	0.109	0.108
78	0.108	0.107	0.107	0.106	0.106	0.105	0.104	0.104	0.103	0.103
79	0.102	0.102	0.101	0.101	0.100	0.100	0.099	0.099	0.098	0.097
	0.0	0.1	0.2	0.3	0.4	0.5	0.6	0.7	0.8	0.9

	0.0	0.1	0.2	0.3	0.4	0.5	0.6	0.7	0.8	0.9
80	0.097	0.096	0.096	0.095	0.095	0.094	0.094	0.093	0.092	0.092
81	0.091	0.091	0.090	0.090	0.089	0.089	0.088	0.088	0.087	0.087
82	0.086	0.086	0.085	0.084	0.084	0.083	0.083	0.082	0.082	0.081
83	0.081	0.080	0.080	0.079	0.079	0.078	0.078	0.077	0.077	0.076
84	0.076	0.075	0.075	0.074	0.074	0.073	0.073	0.072	0.072	0.071
85	0.071	0.070	0.070	0.069	0.069	0.068	0.068	0.067	0.066	0.066
86	0.065	0.065	0.064	0.064	0.063	0.063	0.062	0.062	0.061	0.061
87	0.060	0.060	0.059	0.059	0.058	0.058	0.057	0.057	0.056	0.056
88	0.055	0.055	0.054	0.054	0.053	0.053	0.052	0.052	0.052	0.051
89	0.051	0.050	0.050	0.049	0.049	0.048	0.048	0.047	0.047	0.046
90	0.046	0.045	0.045	0.044	0.044	0.043	0.043	0.042	0.042	0.041
91	0.041	0.040	0.040	0.039	0.039	0.038	0.038	0.038	0.037	0.037
92	0.036	0.036	0.035	0.035	0.034	0.034	0.033	0.033	0.032	0.032
93	0.031	0.031	0.030	0.030	0.030	0.029	0.029	0.028	0.028	0.027
94	0.027	0.026	0.026	0.025	0.025	0.025	0.024	0.024	0.023	0.023
95	0.022	0.022	0.021	0.021	0.020	0.020	0.020	0.019	0.019	0.018
96	0.018	0.017	0.017	0.016	0.016	0.015	0.015	0.015	0.014	0.014
97	0.013	0.013	0.012	0.012	0.011	0.011	0.010	0.010	0.010	0.009
98	0.009	0.008	0.008	0.007	0.007	0.007	0.006	0.006	0.005	0.005
99	0.004	0.004	0.003	0.003	0.002	0.002	0.002	0.001	0.001	0.001
	0.0	0.1	0.2	0.3	0.4	0.5	0.6	0.7	0.8	0.9

2.121. Determination of Solidification Point of Water-Insoluble Fatty Acids (Titre)

(Fifth Edition: Methods II.A.2 and II.B.3)

1. SCOPE

This Standard describes a method for the preparation of the water-insoluble fatty acids of oils and fats, and for the determination of their solidification point (titre).

2. FIELD OF APPLICATION

This Standard is applicable to animal and vegetable oils and fats whose water-insoluble fatty acids have a solidification point of 30°C or above.

3. DEFINITION

The *solidification point of the fatty acids* is the temperature determined by the method described as the maximum of a temporary temperature rise during the cooling of the melted fatty acids. If the latent heat is not sufficient to cause a rise in temperature, the temporary interruption of the cooling process is considered as the solidification point.

4. PRINCIPLE

Saponification of the oil or fat by an ethanolic potassium hydroxide solution. After removal of ethanol, dissolution of the soap in water and decomposition of this by sulphuric acid. Washing separated fatty acids with boiling sodium chloride solution, drying and filtration.

Determination of the highest temperature during the temporary temperature rise, or the interruption of the cooling process, when the melted fatty acids are cooled.

5. APPARATUS

5.1. 500-ml and 1000-ml round-bottomed flasks.

5.2. Reflux condenser to fit the flasks (5.1).

5.3. 100-ml and 1000-ml beakers.

5.4. 1000-ml separating funnel.

5.5. Glass test tube, 120 ± 5 mm in length, 27.5 ± 0.5 mm internal diameter, 2.5 ± 0.5 mm wall thickness. This test tube carries a mark at 55 mm from the bottom.

5.6. 600-ml beaker, squat form (125 mm in height, 91 mm external diameter). This beaker should be weighted (e.g. with lead shot) to keep it steady on the bottom of the water bath (5.7).

5.7. Water bath, maintained 20-25°C below the expected solidification point.

5.8. Thermometer, accurately calibrated, graduated in 0.2°C, with a scale covering all or part of the range from 30 to 70°C, according to the sample to be tested.

5.9. Flat circular cork, having a central hole just big enough to support the test tube (5.5) and able to fit on the beaker (5.6).

5.10. Support to which the thermometer (5.8) can be suspended so that it hangs in the centre of the test tube (5.5).

5.11. Boiling chips.

6. REAGENTS

6.1. Sodium sulphate, anhydrous.

6.2. Potassium hydroxide, ethanolic solution. Dissolve 18 g potassium hydroxide in about 20 ml distilled water, and dilute with 150 ml of 95 per cent (V/V) ethanol.

6.3. Sulphuric acid, aqueous solution. Add 1 volume of sulphuric acid (ρ=1.84 g/ml) to 4 volumes of distilled water.

6.4. Sodium chloride, 100 g/l aqueous solution.

7. PREPARATION OF THE SAMPLE

Use about 100 g of the fat, and put it in an oven maintained at about 20°C above its melting point, until totally melted.

8. PROCEDURE

8.1. Preparation of the Water-Insoluble Fatty Acids

Weigh about 50 g of the prepared sample into a 500-ml flask (5.1). Add all the ethanolic potassium hydroxide solution (6.2) and some boiling chips (5.11). Fit the reflux condenser (5.2). Bring gently to the boil with occasional shaking. Maintain a gentle ebullition during 1 hour. Distil off most of the ethanol (about 125 ml); take care that the soap does not become dry. Add about 250 ml of boiling water and mix to dissolve the soap. Transfer the soap solution into a 1000-ml beaker (5.3), add 500 ml of boiling water, and boil for 45 minutes in order to remove the remainder of the ethanol. Make up the contents to about 700 ml with water.

Pour the soap solution into a 1000-ml flask (5.1) and add carefully 70 ml of the aqueous sulphuric acid solution (6.3). Fit the reflux condenser (5.2) and heat to boiling. Continue boiling until the free fatty acids float to the surface in a clear layer.

Transfer the contents to a separating funnel (5.4) and draw off the aqueous layer as completely as possible.

Wash the fatty acids twice, each time using about 500 ml of aqueous sodium chloride solution (6.4). After each washing draw off the aqueous layer as completely as possible.

Transfer the fatty acids to a 100-ml beaker (5.3). Add a small quantity of anhydrous sodium sulphate (6.1), stir, and filter through a dry filter paper.

Let the insoluble fatty acids crystallize in a desiccator for about 24 hours.

8.2. Determination of the Solidification Point of the Water-Insoluble Fatty Acids

Pour the fatty acids obtained according to 8.1 and melted at a temperature of 25° C above the expected solidification point into a test tube (5.5) up to the 55-mm mark.

Support the tube (5.5) in the cork (5.9) so that a length of about 3 cm projects above it, rest the cork laying on the rim of a beaker (5.6) and place this in a water bath (5.7), so that the water level is 10 mm above the 55-mm mark of the tube. Suspend the thermometer (5.8) in the centre of the tube so that the base of the bulb is 1 cm from the bottom of the tube.

The mercury column falls rapidly at first, and then more slowly. Simultaneously, the fatty acids crystallize, first at the bottom of the tube, and then gradually covering the base of the thermometer bulb.

When the mercury column appears to be stationary during four observations made at 5-second intervals, stir the fatty acids with a rapid, circular movement of the thermometer three times to the right and three times to the left, thus breaking up the crystals formed.

Replace the thermometer immediately in the centre of the tube and take further readings. The mercury column, which falls sharply during the agitation, now rises again and attains a maximum, or remains steady, before falling again. This maximum or standstill is the solidification point.

Carry out two determinations on the same prepared sample.

9. EXPRESSION OF RESULTS

Take as the result the arithmetic mean of the two determinations, if the requirements of repeatability are satisfied. If the difference is greater, repeat the determination until two results differing by not more than 0.2°C are obtained.

10. REPEATABILITY

The difference between the results of duplicate determinations carried out simultaneously or in rapid succession by the same analyst should not exceed 0.2°C.

2.131. Determination of the Bömer Value (B.V.)

(Fifth Edition: Method II.D.18)

1. SCOPE

This Standard describes a test intended to reveal the presence of foreign fat (including hydrogenated or interesterified pork fat) in pork fats (primarily lard) (*note 1*).

The test is based on the fact that, owing to the particular glyceride composition of pork fat, the difference between the melting point of the saturated triglycerides present and the melting point of the fatty acids derived from these triglycerides is greater for pork fat than for the other fats.

2. FIELD OF APPLICATION

This Standard is applicable to pork fats (primarily lard) which are liable to be admixed with foreign fats, which may include hydrogenated or interesterified pork fats.

3. DEFINITION

The Bömer value is defined as the sum of the melting point of the triglycerides isolated by the method prescribed and double the difference between this melting point and that of the corresponding fatty acids.

4. PRINCIPLE

The saturated triglycerides of the fat sample are isolated by recrystallization from a solvent, diethyl ether (Method I) or acetone (Method II).

Part of the triglycerides so recovered are saponified, and the fatty acids recovered by acidification and extraction with a solvent, diethyl ether (Method I) or *n*-hexane (Method II).

The melting points of the saturated triglycerides and of the corresponding fatty acids are determined simultaneously.

5. APPARATUS

5.1. 150-ml (Method I) and 100-ml beakers.

5.2. 60-mm diameter watchglasses.

5.3. Glass flasks with ground glass stoppers.

5.4. 100-ml conical flask, with ground neck.

5.5. Reflux condenser to fit the conical flask (5.4).

5.6. 250-ml separating funnel.

5.7. Glass capillary tubes, sealed at one end; 0.8-1.0 mm internal diameter, 0.1-0.2 mm wall thickness, and 70-80 mm in length.

5.8. 100-ml centrifuge tube, or glass-stoppered cylinder of the same capacity (Method II).

5.9. Buchner funnel of suitable diameter (usually 50 mm).

5.10. Mortar, capacity about 50 ml, preferably of agate.

5.11. Suitable melting point apparatus, filled with a liquid medium and equipped with a stirrer.

5.12. Precision thermometer graduated in 0.2°C from 35° to 85°C.

5.13. Reading lens.

5.14. Water bath, regulated at 15 ± 1°C (Method I) or at 30 ± 2°C (Method II).

5.15. Water bath.

5.16. Oven regulated at 35°C.

5.17. Oven regulated at 103 ± 2°C.

5.18. Desiccator, containing conc. sulphuric acid or an equally effective acid desiccant.

5.19. Filter paper, fast-filtering type.

6. REAGENTS

6.1. *n*-Hexane.

6.2. Ethanol, 95 per cent (V/V).

6.3. Diethyl ether, free from peroxides and recently distilled (Method I).

6.4. Acetone (Method II).

6.5. Potassium hydroxide, pellets.

6.6. Sodium sulphate, anhydrous.

6.7. Sulphuric acid, approx. 1 N aqueous solution.

6.8. Filter aid.

7. PREPARATION OF THE SAMPLE

Melt the sample of fat at about 10°C above its melting point and mix well. At this temperature take about 150 g of the molten fat, add 4 g of anhydrous sodium sulphate (6.6) and 1 g of filter aid (6.8) and filter through a dry, fast-filtering paper (5.19). The filtrate must be perfectly clear.

8. PROCEDURE: METHOD I WITH DIETHYL ETHER

8.1. Separation of the Triglycerides

Into a 150-ml beaker (5.1) weigh about 50 g of the prepared sample. Add 50 ml of diethyl ether (6.3), cover with a watchglass (5.2) and dissolve the fat by swirling and gently heating the beaker over a water bath. The solution must be clear; if not, filter through a dry filter paper (5.19).

Place the beaker, covered with the watchglass, in a water bath maintained at 15 ± 1°C (5.14) and leave it there for 1 hour, mixing the contents every 15 minutes by swirling or by stirring with a glass rod, without removing the beaker from the bath.

If no crystals have formed after 30 minutes, continue the crystallization for the next 30 minutes at a lower temperature, but above 5°C. If under these conditions less than 0.5 g of crystals are obtained, repeat the whole operation with a further test portion and combine the two crops of crystals.

Filter off the crystals, if necessary under low vacuum, on a Buchner funnel (5.9) (of diameter appropriate to the volume of the crystals) fitted with a fast-filtering paper (5.19), previously wetted with diethyl ether. Immediately wash the crystals on the funnel three times with 25-ml portions of diethyl ether cooled to 15°C.

Detach the crystals from the filter paper and return them to the 150-ml beaker which was used for the crystallization after first rinsing it with diethyl ether. Add 50 ml of diethyl ether, cover with a watchglass, dissolve the crystals by swirling or by stirring with a glass rod, and gently heating the beaker over a water bath.

Place the beaker covered with the watchglass in a water bath maintained at 15 ± 1°C for 15 minut

Filter off the crystals, if necessary under low vacuum, on a Buchner funnel fitted with a fast-filtering paper previously wetted with diethyl ether. Immediately wash the crystals on the funnel three times with 25-ml portions of diethyl ether cooled to 15°C.

Detach the crystals from the filter paper and return them to the 150-ml beaker which was used for the crystallization after first rinsing it with diethyl ether. Add 50 ml of diethyl ether, cover with a watchglass, dissolve the crystals by swirling or by stirring with a glass rod, and gently heating the beaker over a water bath.

Place the beaker covered with the watchglass in a water bath maintained at 15 ± 1°C for 15 minutes.

Place the filter paper and crystals from the third crystallization on a watchglass and dry in the oven (5.16) at a temperature not exceeding 35°C.

Grind the crystals to a fine smooth powder in a mortar (5.10). Store these powdered saturated triglycerides in a glass bottle (5.3) provided with a ground stopper in a desiccator (5.18).

8.2. Preparation of the Fatty Acids

Warning. The preparation of the fatty acids must be performed in an atmosphere free from ammonia.

Place 0.2 g (*note 2*) of the powdered saturated triglycerides obtained as above into a 100-ml conical flask (5.4). Add 10 ml of ethanol (6.2) and 0.4 g (2 pellets) of potassium hydroxide (6.5). Fit the reflux condenser (5.5) and boil gently for 15 minutes.

Remove the condenser and pour the ethanolic soap solution into a 100-ml beaker (5.1).

Place the beaker in a boiling-water bath (5.15) in order to evaporate the bulk of the ethanol.

Add 50 ml of water at about 75°C to dissolve the soap and transfer the solution to a separating funnel (5.6).

While stirring, add 10 ml of the aqueous sulphuric solution (6.7) and allow to cool to room temperature.

When cool add 50 ml of diethyl ether (6.3), shake to dissolve the liberated fatty acids, and allow to settle.

Remove the aqueous layer. Wash the solution of the fatty acids with 15-ml portions of distilled water at least three times, or more if necessary, until the wash-water is neutral to methyl orange; take care to remove the final wash-water completely.

Add about 4 g of anhydrous sodium sulphate (6.6) to the solution of the fatty acids and filter through a dry filter paper into a 100-ml beaker.

Evaporate the solvent on the water bath (5.15).

Place the beaker in a drying oven (5.17) maintained at 103 ± 2°C and leave it there for 15 to 20 minutes.

Remove the beaker from the oven and allow to cool, keeping it in an inclined position.

Detach the fatty acid cake and pulverize it in a mortar (5.10) to a fine, smooth powder. The fineness of the powder should be substantially the same as that of the saturated triglyceride powder.

Store the powdered fatty acids in a glass bottle (5.3) in a desiccator (5.18).

8.3. Determination of the Melting Points

Fill a capillary tube (5.7) with the powdered saturated glycerides. Pack the material to a height of about 5 mm, for instance, by letting the capillary tube drop on to a wooden board down a glass tube of sufficient height (30 to 40 cm). Never melt the glycerides in an attempt to fill the capillary.

Fill another capillary tube in the same way with the powdered fatty acids as soon as possible after their preparation. Melt the contents of this tube at 70°C for at least 1 minute, but not exceeding 5 minutes, and allow to solidify by cooling.

Fit the two capillary tubes and the thermometer (5.12) into the melting point apparatus (5.11).

If necessary, fix the two tubes to the thermometer with the help of a rubber band, so that the centres of the thermometer bulb and of the sample columns are on the same level. Raise the temperature at the rate of 2°C per minute up to 50°C, and after at a rate of 0.5°C per minute.

With the help of the lens (5.13), observe the temperature at which the last (solid) particle vanishes in each tube. Record these temperatures to within 0.1°C as the respective melting points.

Carry out two complete determinations on the same prepared sample.

9. PROCEDURE: METHOD II WITH ACETONE

9.1. Separation of the Triglycerides

Into a 100-ml centrifuge tube or stoppered cylinder (5.8), weigh about 20 g of the prepared sample.

Add 80 ml of acetone (6.4) at 30°C, insert the stopper and shake until thoroughly mixed. Stand the tube in a water bath (5.14) at 30 ± 2°C for about 18 hours.

If the amount of crystals is less than 0.5 g, repeat the operation with another test portion, and combine the two crops of crystals.

Separate the crystals obtained, either by centrifugation (for 5 minutes) and decanting the supernatant liquid, or by siphoning off the supernatant liquid if a cylinder was used.

Add 20 ml of acetone at 30°C, shake and separate the crystals as before.

Again add 20 ml of acetone at 30°C to the crystals; mix well, and filter under slight suction through a filtering paper (5.19) on a Buchner funnel (5.9) (the size depending on the amount of crystals to be separated).

Wash the crystals on the filter five times with 5-ml portions of acetone at 30°C, the last wash being made under partial vacuum.

Place the filter paper with the crystals on a watchglass (5.2) and dry in an oven (5.16) at a temperature not exceeding 35°C.

Transfer the crystals to a mortar (5.10), break up any lumps and allow to dry thoroughly; continue drying for 15 minutes at a temperature not exceeding 35°C. Grind the crystals in the mortar (5.10) to a fine smooth powder. Store this triglyceride powder in a glass bottle (5.3) in a desiccator (5.18).

9.2. Preparation of the Fatty Acids

Proceed exactly as in 8.2, substituting *n*-hexane (6.1) for diethyl ether (6.3).

9.3 Determination of the Melting Points

Proceed exactly as in 8.3.

10. EXPRESSION OF RESULTS

The Bömer value (B.V.) is given by the formula:

$$B.V. = t_g + 2\ (t_g - t_a)$$

where t_g is the melting point of the powdered saturated triglycerides in °C,
t_a is the melting point of the powdered fatty acids in °C.

11. REPEATABILITY

The difference between the results of two complete determinations of B.V., made simultaneously or in rapid succession by the same analyst, may not be greater than 0.5. If this limit is exceeding, the whole operation must be repeated.

12. REMARKS

The two methods described yield practically identical results. The method employing diethyl ether is to be preferred, however, because it has a smaller standard deviation. The acetone method is to be used if climatic or legal considerations limit the use of diethyl ether.

13. NOTES

1. In view of the particular nature of the methods described, it is essential to adhere scrupulously to the operational details prescribed in order to procure reproductible results.

2. The amount of the powdered saturated triglycerides must be such that, after taking 0.2 g, there will remain sufficient for the determination of the melting point. For it is essential that the saponification shall always be performed under identical conditions.

2.132. Determination of the Cooling Curve of Fats

(Fifth Edition: Method II.B.5)

1. SCOPE

This Standard describes a method for the determination of the cooling curve of fats.

2. FIELD OF APPLICATION

This Standard is applicable to animal and vegetable fats.

3. PRINCIPLE

In an apparatus conforming strictly to the specifications given, cooling of the fat in an ice-water mixture or a water bath at 17°C. According to the precise conditions of the method, determination of the temperature of the sample at regular time intervals. Construction of the graph relating temperature to time and interpretation of results.

Since the phase transitions are accompanied by thermal changes, they may be indicated by changes in the slope of the curve. Various fats (especially cocoa butter) and mixtures of fats give characteristic cooling curves.

As the method is empirical, it is essential to use apparatus conforming strictly to the specifications given.

4. APPARATUS

4.1. Glass cooling flask conforming to the diagram, where all dimensions are given in mm. This modified Shukoff flask, which has a capacity of about 30 ml, is vacuum jacketed; the space between the walls is evacuated to a pressure (usually about 0.01 mm Hg) such that when the cooling curve for a freshly refined soyabean oil is determined according to the method described in 5.1 and/or 5.2 the following figures (with a tolerance of ± 0.3°C) are obtained:

Time (min)	Temperature (°C) Method 5.1	Temperature (°C) Method 5.2
0	40.0	40.0
5	32.8	35.4
10	27.2	31.7
15	22.4	28.9
20	18.5	26.5

4.2. Thermometer, range 0 to 50°C, graduated in 0.1°C; recommended dimensions: total length 40 cm, bulb diameter 6 mm, bulb length 12 mm; the thermometer is fitted into the flask by means of a ground-glass stopper or rubber stopper so that the bulb is *exactly in the centre* of the flask.

4.3. 1-litre Dewar flask, filled with an ice-water mixture at 0°C.

4.4. Water bath regulated at 17 ± 0.1°C.

4.5. Stop-watch.

5. PROCEDURE

N.B. The shape of the cooling curve may vary if oxidation products are present in the sample. Consequently, it is necessary for the test sample to be fresh, and repeated heating must be avoided.

Heat about 30 g of the sample to 70°C or, if necessary, at a higher temperature, until all crystals are completely melted. Filter carefully through a dry filter paper at a temperature at least 20°C above the melting point so as to obtain a perfectly clear liquid.

Introduce 25 ± 0.2 g of the melted sample into the flask (4.1) and fit the thermometer (4.2) (*note 1*) when the temperature reaches 50°C. Shake the flask so as to bring the whole of the contents to a uniform temperature, and immerse the flask in the cooling bath as described below.

5.1. Cooling Bath at 0°C

Immerse the flask in the ice-water mixture in the Dewar flask (4.3) to the level indicated in the diagram. When the temperature of the test portion reaches 40.0°C, start the stop-watch and read the thermometer every minute; construct the time-temperature graph.

5.2. Cooling Bath at 17°C (*note 2*)

Immerse the flask in the water bath (4.4) maintained at 17 ± 0.1°C to the level indicated in the diagram, and proceed as described in 5.1.

6. INTERPRETATION OF THE COOLING CURVE

Compare the cooling curve obtained with the sample with the reference curve obtained from soyabean oil (cf. 4.1).

6.1. Prime Stay Temperature

The two curves coincide at the beginning of the cooling process. But at a certain temperature they begin to diverge, the temperature of the sample being above that of the oil, owing to the onset of crystallization. Record this temperature, the so-called "prime stay temperature", and the corresponding time (*note 3*).

6.2. Minimum and Maximum Temperatures

Following the prime stay temperature, the cooling curve of most fats shows a minimum followed by a rise to a maximum peak. Record these temperatures and the corresponding times. Certain fats, however, may only show a rather flat wave with an ill-defined minimum or maximum, or, in some cases, only a regularly falling curve.

7. EXPRESSION OF RESULTS

From the graph obtained report the following:

(a) temperature of the cooling bath,

(b) temperature and time of the point of prime stay,

(c) temperature and time at the minimum point on the graph,

(d) temperature and time at the peak point on the graph,

(e) differences in temperature and in time between the minimum and peak points on the curve.

8. NOTES

1. If the apparatus conforms to the specification, the level of the sample will be as indicated in the diagram.

The amount of oil used is less important than the degree of vacuum in the jacket, provided that, when the flask is filled, the level of the oil is at least 20 mm below the upper rim of the jacket.

2. This method is suitable for cocoa butter and cocoa butter substitutes, but, in the case of certain fats, such as coconut and palmkernel oils, a temperature of 17°C for the cooling bath may be too high; in such cases it is preferable to use the procedure 5.1.

3. If, after some minutes, the two curves do not coincide exactly, because the soyabean oil shows temperature readings slightly below those of sample, then plot a graph showing the variation of the difference between the temperatures of the sample and soyabean oil as a function of time. The time at which this graph shows a change in slope is taken as the prime stay temperature.

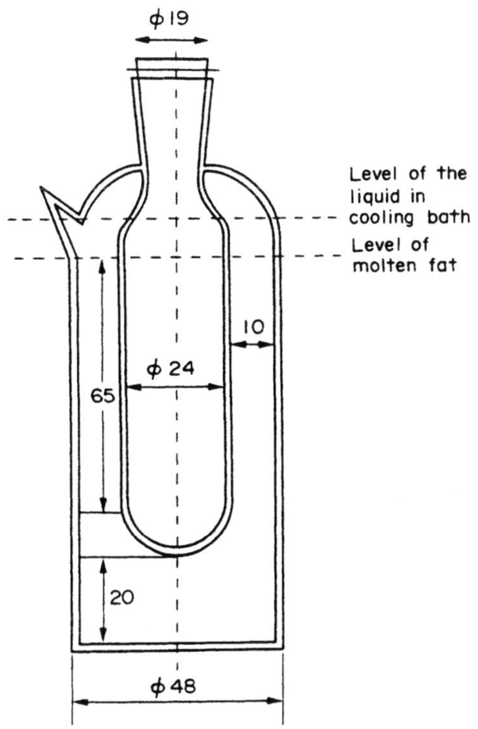

Cooling curve flask (dimensions in millimetres).

2.141. Determination of the Dilatation of Fats

(Fifth Edition: Method II.B.6)

1. SCOPE

This Standard describes a method for the determination of the dilatation of fats.

2. FIELD OF APPLICATION

This Standard is applicable to vegetable and animal fats having a melting point above 0°C. The method is empirical and therefore any variation in the prescribed operational details will modify the results obtained.

A special procedure is given for the stabilization of cocoa butter and similar fats.

3. DEFINITION

The dilatation of a fat is the isothermal expansion, due to change of state from solid to liquid, of the fat which has previously been solidified under precisely prescribed conditions.

The dilatation is expressed in *ml per kg of fat* (µl per g) and represents the difference between the volume the fat would occupy at the temperature in question if it were completely liquid (supercooled) and the actual volume of the fat at this temperature. It affords an empirical indication of the amount of crystalline matter in the fat at this temperature.

N.B. The dilatation is sometimes required to be expressed in *µl per 25 g of fat*; in such case, the value for the dilatation obtained by the method described must be multiplied by 25.

4. PRINCIPLE

The volume of a known weight of the fat is measured at various temperatures below 60° C. The volume of the liquefied fat is measured at 60°C; the volumes of the liquid supercooled fat at the other required temperatures are derived by calculation from the volume at 60°C.

5. APPARATUS

5.1. 100-ml round-bottomed flask.

5.2. Burette graduated in 0.05 ml, or smaller divisions.

5.3. Dilatometers, as illustrated in the diagram.

The bulb of the dilatometer should have a volume of 7.0 ± 0.5 ml, and the graduated part of the capillary tube a volume of 900 µl. The zero mark on the capillary tube should be level with the top of the neck of the dilatometer bulb. The stopper should be filled with sufficient lead shot to prevent the dilatometer floating in the water bath.

5.4. Constant temperature water baths.

Each bath must be fitted with:

(a) a thermostat capable of maintaining the water at any selected temperature between 10°C and 60°C to within ± 0.1°C,

(b) a vigorous stirrer,

(c) a thermometer (5.6), accurate to within 0.05°C.

If only one water bath is used, the heating elements must have sufficient capacity to raise the temperature of the water by 5°C within 5 minutes at most. If two water baths are used, the dilatometers are transferred into the bath adjusted to the next temperature at which the dilatation is to be determined. This implies that the heating capacity of each bath should be sufficient to raise the temperature of the water by 10°C within a maximum of 30 minutes.

The baths must be equipped with an arrangement for suspending the dilatometers in the water, which allows individual vertical adjustment and easy transfer from one bath to another.

When meniscus levels are read, the temperature of the water in the immediate neighbourhood of the dilatometer should not deviate by more than 0.05°C from the average temperature of the bath (*note 1*).

5.5. Bath at 0°C.

If crushed ice is used for the bath at 0°C, the vessel must have draining holes, otherwise water at a temperature above 0°C will collect at the bottom of the vessel. The ice must frequently be stirred and repacked, especially if several dilatometers are being cooled from 60°C to 0°C in the same bath. A cooling unit that can be maintained at 0°C is recommended, but this will need to be filled with a liquid having a freezing point below 0°C.

5.6. Thermometers graduated in 0.1°C (or 0.05°C), covering the required range of temperatures; they must be calibrated against an official standard thermometer.

5.7. Water bath.

5.8. Boiling chips.

6. TESTING OF THE DILATOMETER CAPILLARY

6.1. Regularity of the Graduated Scale

Introduce through the bulb into the capillary of the carefully cleaned dilatometer (5.3) approximately 250 µl of mercury. Place the dilatometer in a horizontal position on two wooden blocks and place a small mirror under the capillary to avoid errors of reading due to parallax. When the dilatometer and mercury have attained room temperature, read the volume of mercury in scale units at various positions along the capillary (*note 2*).

The differences between the readings must not exceed 2 µl. If they do, reject the capillary. A regular increase, or decrease, in the values along the capillary is not acceptable, and the tube must be rejected.

6.2. Accuracy of the Graduated Scale

Next, empty the capillary and fill it again via the bulb with about 750 µl of mercury. Place the dilatometer in a horizontal position on the wooden blocks alongside a thermometer, and arrange the mirror under the capillary.

Read the volume of the mercury several times after slight displacement of the mercury column and note the temperature (*note 2*).

Weigh the mercury in the capillary tube after pouring it out from the 900-µl-graduated end of the tube into a tared flask.

Calculate the volume of the mercury from its mass and apparent density at the measuring temperature. Deduct 1 µl from the observed volume as a correction for reading the meniscus, and compare the corrected observed volume with the true volume.

Reject dilatometers showing deviations of 0.5 per cent or more.

7. GENERAL METHOD

7.1. Scope

This method is not applicable to cocoa butter and similar fats.

7.2. Procedure

Three dilatation procedures may be used :

(a) *normal short range :* temperatures of measurement : 20°C, 25°C, 30°C, 35°C and every 5°C up to 60°C if required;

(b) *normal long range :* temperatures of measurement : 0°C, 10°C, 15°C, 20°C, 25°C, and every 5°C up to 60°C if required;

(c) *rapid method for factory control* : temperatures of measurement : 20°C and 30°C, and any higher temperature if required.

If the melted fat is not clear, filter it. Transfer 20 ml or less of the fat into a flask (5.1), add a small boiling chip (5.8), heat to 80-100°C **in a water bath** (5.7) and evacuate the flask in a sufficient vacuum, shaking vigorously at intervals until air bubbles no longer escape from the boiling chip. This may take 10 minutes (*note 3*).

Keep the melted fat under vacuum, and keep it warm (70°C) until it has been put into the dilatometer (5.3). By means of a burette (5.2) put exactly 1.5 ml of sealing liquid, consisting of distilled water boiled out under vacuum and coloured with dye (congo red, methyl orange or 1 per cent potassium dichromate) into the dilatometer. Weigh the dilatometer with the stopper to the nearest 10 mg.

Fill the dilatometer bulb by halfway up the neck with the prepared liquid fat (*note 4*). Insert the stopper carefully to avoid trapping air-bubbles and at the same time to drive the sealing liquid up in the capillary. If the sealing liquid in the capillary rises to 600 - 700 μl, the capillary should be closed almost completely with the tip of the finger; when the stopper is securely in position, remove the finger. Check that no air has been trapped under the stopper. Clean the outside of the bulb with solvent, dry and weigh the dilatometer again to determine the mass of the fat.

Immerse the dilatometer to the zero mark in the water bath (5.4) maintained at 60°C ± 0.1°C. Read the level of the sealing liquid in the capillary after not less than 30 minutes. A burette magnifier increases the precision with which the reading can be made.

Remove the dilatometer from the water bath and immerse it to the zero mark in crushed ice or in a bath at 0°C (5.5) for 90 minutes. Note the level of the meniscus in the capillary if a value D_0 is required.

Remove the dilatometer from the 0°C bath and immerse it to the zero mark in a water bath (5.4) maintained at the lowest temperature at which the dilatation is to be measured. Read the level of the meniscus 30 minutes after the water bath has again reached the required measuring temperature. All the dilatometers in use should be read within 5 minutes.

Take readings of the meniscus level similarly at successively higher temperatures 30 minutes after the water bath has attained the measuring temperature in each case.

Always repeat the reading at 60°C. This value should agree to within 2 μl with the value found at the start. A greater difference could be a consequence of a leak in the dilatometer or of the presence of air in the fat.

If, by mistake, the temperature of a bath should exceed the intended temperature, the bath should not be cooled, because the melted part of the fat will not recrystallize. Maintain the wrong temperature and report the dilatation at this temperature.

When reporting the results, the designation "general method" must be included, as well as the designation "short range", "long range" or "rapid method", since the numbers of temperature steps employed influences the dilatation values.

8. METHOD FOR COCOA BUTTER AND SIMILAR FATS

8.1. Foreword

Unhardened fats which consist principally of 2-oleo 1-3-disaturated triglycerides, display a characteristic polymorphism and must be stabilized in a particular way before the dilatations are measured. Nevertheless, for cocoa butter-like fats substitutes based on lauric acid derivatives, whether hydrogenated or not, the general method 7 may be applied. It should then be stated in the Report whether the fat has been subjected to a pre-treatment or not.

Fats very rich in 2-oleo 1-3-distearide, such as shea fat fractions, kokum butter, and the *Allanblackia* group of fats sometimes behave unsatisfactorily if occluded air is retained; it is then advisable to heat the dilatometer before filling it.

8.2. Procedure

Temperatures of measurements : 20°C, 25°C, 30°C, 35°C and every 5°C up to 60°C if required.

Proceed as in 7.2 from "If the melted fat is not clear" to ..."for 90 minutes".

Transfer the dilatometer to a water bath (5.4) fitted with a stirrer and maintained at 26 ± 0.2°C keep it there for 40 ± ½ hours. During this time, the capillary tube must be capped in order to avoid evaporation of the sealing liquid.

Immerse the dilatometer again to the zero mark in crushed ice or in a bath at 0°C (5.5) for 90 minutes. Note the level of the meniscus in the capillary tube if the value D_0 is required, then place the dilatometer in a water bath (5.4) maintained at the lowest temperature (20°C) of the series for which dilatation is to be measured. Read the level of the meniscus 60 minutes after the water bath has recovered the required temperature; all the dilatometers in use must be read within 5 minutes.

Proceed as in 7.2 from "Take readings of the meniscus similarly" up to "report the dilatation at that temperature".

The designation "Method for cocoa butter" should be included in the test report.

9. EXPRESSION OF RESULTS

The dilatation D_t of the fat at temperature t°C is given by the formula :

$$D_t = \frac{A_{60} - A_t - W_t}{m} - V_t$$

where m is the mass, in g, of the fat in the dilatometer,
A_{60} is the reading at 60°C, in µl,
A_t is the reading at t°C, in µl,
W_t is the correction for the sealing liquid, i.e. the correction for the expansion of sealing liquid and glass between 60°C and t°C (see Table 1),
V_t is the expansion of the oil, i.e. the difference between the volumes (in µl) of 1 g of melted fat at 60°C and of supercooled fat at t°C (see Table 2).

The dilatation values should be rounded off to the nearest half unit.

A model form of report for the calculation of the dilatation is shown in the Appendix; the use of this form is recommended.

Table 1

To find the correction W_t in µl (Note 5) for the expansion of water (used as the sealing liquid) and glass at various temperatures and various observed levels of the meniscus (from the zero of the capillary tube to the level of the water in the water bath) round off the height of the meniscus to multiples of 100, and read the value of W_t from the table:

Level of the meniscus in µl

t°C	0	100	200	300	400	500	600	700	800	900
0	24	22	20	18	16	14	12	11	9	7
10	22	21	19	17	15	14	12	10	9	7
15	21	20	18	16	15	13	12	10	8	7
20	20	18	17	15	14	12	11	9	8	6
25	18	17	16	14	13	11	10	9	7	6
30	17	15	14	13	12	10	9	8	6	5
35	15	13	12	11	10	9	8	7	6	5
40	12	11	10	9	9	8	7	6	5	4
45	10	9	8	7	7	6	5	5	4	3
50	7	6	6	5	5	4	4	3	3	2
55	3	3	3	3	2	2	2	2	1	1
60	0	0	0	0	0	0	0	0	0	0

Table 2

V_t (*note 5*), i.e. the expansion of the oil between t°C and 60°C :

Temperature (°C)	V_t in µl per g
0	50.6
10	42.4
15	38.3
20	34.1
25	29.9
30	25.7
35	21.5
40	17.2
45	13.0
50	8.6
55	4.3
60	0

10. NOTES

1. The efficiency of the stirring and possible deviations of the temperature of the water in the immediate neighbourhood of the dilatometer (5.3) can be checked by inserting a dilatometer filled with a hydrogenated fat, e.g. hydrogenated palm oil, and fitted with a thermometer (5.6).

The reading of this thermometer should not differ by more than 0.05°C from those of the other thermometer in the water bath.

2. During the reading of the mercury meniscus in testing the capillary tubes, avoid heating the tubes by the hands or by breathing upon them.

3. Deaeration of the sample

An alternative method may be used : place 60 ml of fat in a 300-ml Kjeldahl flask. Close the flask with a rubber stopper carrying a vacuum lead and a thermometer (200°C) which reaches into the fat. Heat the combination intermittently over a Bunsen flame, so as to maintain the oil at 110°C, swirling continually and shaking vigorously at intervals, until all bubbling due to air or moisture in the fat has ceased. Avoid any over-heating of the fat.

4. In the case of fats with a high melting point, or which crystallize rapidly, the dilatometer must be warmed in order to prevent solidification of the fat.

5. Corrections for the expansion of glass, of the sealing liquid and of the oil

(a) *Glass correction :* for a 7-ml dilatometer the glass correction, based on a coefficient of linear expansion of 33×10^{-7} cm/°C, is 0.07 µl/°C.

(b) *Correction for the expansion of water and glass* between t°C and 60°C :

The equation giving the variation in the coefficient of expansion of water between 0°C and 60°C may be taken as linear. Using the known values above 20°C and extrapolating to 0°C the equation becomes :

$$E_t = (60 - t)\,\frac{2.3 + 0.09\,(60 + t)}{25}$$

This quantity must be multiplied by the volume of water remaining in the bulb at each temperature (1.5 − 0.001 x reading at t°C) to give the water correction. The mean glass correction, i.e. 0.07 (60 − t), must be substracted from this to yield the figures shown in Table 1.

(c) *Correction for the expansion of the oil :* The correction V_t (in µl/g) for the expansion of the oil is given by the following equation :

$$V_t = \frac{1}{25}\int_t^{60} (20.5 + 0.02t)\,dt$$

APPENDIX

<u>D I L A T A T I O N</u> Report

SAMPLE : .... DATE : ....

DILATOMETER NO. : ....

PROCEDURE : General method / Cocoa butter method *

TEMPERATURE RANGE : Short range / Long range / Rapid method *

MASS OF THE DILATOMETER with SEALING LIQUID and OIL : ....g

MASS OF THE DILATOMETER with SEALING LIQUID : ....g

MASS OF OIL = m : ....g

* Delete whichever is not applicable

		I	II	III	IV	V	D_t
temp. °C	A_t	$A_{60}-A_t$	$W_t^{(1)}$	I - II	$\frac{III}{m}$	$V_t^{(2)}$	IV - V
60						0	
0						50.6	
10						42.4	
15						38.3	
20						34.1	
25						29.9	
30						25.7	
35						21.5	
40						17.2	
45						13.0	
50						8.6	
55						4.3	
60						0	

(1) See Table 1.

(2) See Table 2.

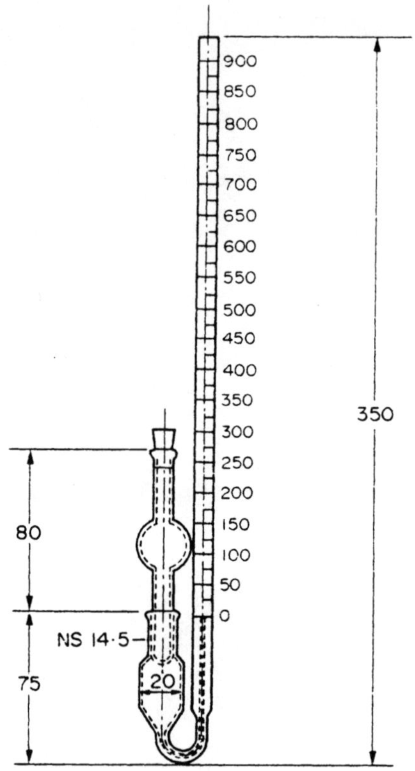

Dilatometer (dimensions in millimetres).

Section II

Oils and Fats

Determination of the Chemical Characteristics

2.201. Determination of the Acid Value (A.V.) and the Acidity

(Fifth Edition : Method II.D.1)

1. SCOPE

This Standard describes two methods for the determination of the free fatty acid content of animal and vegetable oils and fats. This can be expressed as an acid value, or as acidity.

2. FIELD OF APPLICATION

This Standard is applicable to animal and vegetable oils and fats. It is not applicable to waxes.

3. DEFINITIONS

The acid value (A.V.) is the number of mg of potassium hydroxide required to neutralize the free fatty acids in 1 g of the fat.

The acidity is a conventional expression of the percentage of free fatty acids. According to the nature of the fat it can be expressed as :

Nature of fat	Expressed as	Molecular weight
Coconut, palm kernel and similar oils	Lauric acid	200
Palm oil	Palmitic acid	256
All other oils	Oleic acid	282

When the result is reported as "acidity", but not further defined, then conventionally the acidity is always expressed as oleic acid.

If the sample contains mineral acids, by convention they are calculated as fatty acids.

4. ACID VALUE — INDICATOR METHOD

4.1. Principle

Solution of a known quantity of the fat to be analysed in a mixture of ethanol and diethyl ether, followed by titration of the free fatty acids present with an ethanolic solution of potassium hydroxide.

4.2. Apparatus

4.2.1. 250-ml conical flask.

4.2.2. 25-ml burette, graduated in 0.1 ml.

4.3. Reagents

4.3.1. Solvent mixture 1/1 (V/V) of 95 per cent (V/V) ethanol and diethyl ether. Neutralize exactly, just before use, by means of the potassium hydroxide solution (4.3.2) in the presence of 0.3 ml of phenolphthalein solution (4.3.3) per 100 ml of the mixture (*note 1*).

4.3.2. Potassium hydroxide, 0.1 N, or if necessary 0.5 N, solution in ethanol, accurately standardized.
The exact concentration should be known, and checked immediately before use. Use a solution prepared at least 5 days previously and decanted into a bottle of brown glass, provided with a rubber stopper. The solution should be colourless or straw yellow (*note 2*).

4.3.3. Phenolphthalein, 10 g/l solution in 95 per cent (V/V) ethanol.

4.4. Procedure

Prepare the sample according to 2.001, "Preparation of the sample". Determine size of sample according to the following table:

Expected acid value	Mass of test portion, in g	Weighing accuracy, in g
<1	20	0.05
1 - 4	10	0.02
4 - 15	2.5	0.01
15 - 75	0.5	0.001
>75	0.1	0.0002

Weigh the test portion into a conical flask (4.2.1). Dissolve it in about 150 ml of the solvent mixture (4.3.1). Titrate, with shaking, with the solution of potassium hydroxide (4.3.2) to the end point of the indicator (pink colour of phenolphthalein persisting for at least 10 seconds) (*note 3, note 4*).

If the quantity of 0.1 N potassium hydroxide solution required exceeds 20 ml, a 0.5 N solution should be used (*note 5*).

4.5. Expression of Results

The acid value (A.V.) is given by the formula :

$$A.V. = \frac{56.1 \times T \times V}{m}$$

where V is the number of ml of the standardized potassium hydroxide solution (4.3.2) used (*note 6*),
T is the exact normality of the standardized potassium hydroxide solution (4.3.2),
m is the mass, in g, of the test portion.

5. ACID VALUE — POTENTIOMETRIC METHOD

5.1. Foreword

This method should be employed instead of Method 4 for the determination of free acidity in highly coloured oils and fats.

5.2. Principle

The free fatty acids are titrated potentiometrically in a non-aqueous medium with a solution of potassium hydroxide in *iso*propanol.

5.3. Apparatus

5.3.1. 150-ml tall form beakers.

5.3.2. 1000-ml volumetric flasks.

5.3.3. 25-ml burette, graduated in 0.1 ml.

5.3.4. pH meter equipped with glass and calomel electrodes (*note 7*).
Contact between the saturated potassium chloride solution and the test solution is made across a sintered glass or porcelain disc, at least 0.3 cm thick.

5.3.5. Stirrer, preferably magnetic stirrer.

5.4. Reagents

5.4.1. 2-Propanol (*iso*propanol), analytical reagent quality.

5.4.2. Methylisobutylketone, analytical reagent quality, neutralized immediately before use by the *iso*propanolic potassium hydroxide solution (5.4.4) to a faint pink colour with phenolphthalein indicator.

5.4.3. Benzoic acid, acidimetric standard quality.

5.4.4. Potassium hydroxide, 0.1 N solution in 2-propanol (5.4.1), accurately standardized.

Dissolve 7 g of potassium hydroxide pellets in 1000 ml of 2-propanol (5.4.1).

5.4.5. Potassium hydroxide, 0.5 N solution in 2-propanol, accurately standardized.
Dissolve 35 g of potassium hydroxide pellets in 1000 ml of 2-propanol (5.4.1).

Standardize the solutions as follows : weigh accurately to within 0.0002 g about 0.15 g (for the 0.1 N solution) or 0.75 g (for the 0.5 N solution) of the standard benzoic acid (5.4.3). Transfer into a beaker (5.3.1) and dissolve in 50 ml of methylisobutylketone (5.4.2).
Insert the electrodes of the pH meter (5.3.4.), start the stirrer (5.3.5) and titrate with the potassium hydroxide solution (5.4.4 or 5.4.5) to the equivalence point (*note 8*).

$$\text{Normality} = \frac{1000 \times m_0}{122.1 \times a_0}$$

where a_0 is the number of ml of the potassium hydroxide solution (5.4.4 or 5.4.5) used,
m_0 is the mass, in g, of benzoic acid taken.

5.5. Procedure

Prepare the sample according to 2.001, "Preparation of the sample".

Weigh to within 0.01 g 5 to 10 g of prepared fat into a beaker (5.3.1). Dissolve in 50 ml of methylisobutylketone (5.4.2).

Insert the electrodes of the pH meter (5.3.4) and start the stirrer (5.3.5). Titrate with the potassium hydroxide solution (5.4.4 or 5.4.5) 0.1 or 0.5 N (according to the acidity of the sample) to the equivalence point (*note 8*).

5.6. Expression of Results

The acid value (A.V.) is given by the formula :

$$\text{A.V.} = \frac{56.1 \times T \times N}{m}$$

where V is the number of ml of the standardized *iso*propanolic potassium hydroxide solution (5.4.4 or 5.4.5) used,
T is the exact normality of the standardized *iso*propanolic potassium hydroxide solution (5.4.4 or 5.4.5) used,
m is the mass, in g, of the test portion.

6. ACIDITY

6.1. Principle

The acidity is calculated from the results obtained in the determination of the acid value by the indicator method or the potentiometric method.

6.2. Expression of Results

The acidity, in per cent, is given by the formula :

$$\frac{T \times V \times M}{10\ m}$$

where V is the number of the standardized ethanolic (4.3.2) or *iso*propanolic (5.4.4 or 5.4.5) potassium hydroxide solution used,
T is the exact normality of the standardized ethanolic (4.3.2) or *iso*propanolic (5.4.4 or 5.4.5) potassium hydroxide solution used,
M is the molecular weight adopted (200, 256 or 282 according to the Table in 3),
m is the mass, in g, of the test portion.

7. NOTES

1. In the solvent mixture diethylether eventually may be substituted by toluene.

2. A stable colourless solution of potassium hydroxide can be prepared in the following manner : reflux 1000 ml of ethanol with 8 g of potassium hydroxide and 5 g of aluminium pellets for 1 hour, then distil immediately. Dissolve the required amount of potassium hydroxide in the distillate. Allow the whole to stand for several days and decant the clear supernatant liquid from the deposited potassium carbonate.

The solution can also be prepared without distillation in the following manner : add 4 ml of aluminium butylate to 1000 ml ethanol and allow the mixture to stand for several days. Decant the supernatant liquid and dissolve therein the necessary amount of potassium hydroxide. This solution is ready for use.

3. In the case of very low acid value (<1) it is useful to blow a low stream of nitrogen into the solution to be titrated.

4. The ethanolic standardized potassium hydroxide solution (4.3.2) can be substituted by an aqueous solution when the acid value is low.

5. If the solution becomes cloudy during the titration, add a further quantity of the solvent mixture (4.3.1).

6. The normality of the ethanolic standardized potassium hydroxide solution varies with the temperature and it may be useful to correct as follows:

$$V = V_t\,[1 - 0.0011(t - t_o)]$$

where V_t is the volume at the temperature t,

t is the temperature, in °C, at which the determination is made,

t_o is the temperature, in °C, at which the ethanolic potassium hydroxide solution has been standardized.

7. It is advisable to store the glass electrode in water or, better still, in methylisobutylketone for 12 hours preceding the titration. Dry it very gently with a piece of filter paper before use. Immediately after the titration, rinse it with methylisobutylketone, then with 2-propanol, and finally with distilled water.

If the electrode does not function satisfactorily, it may be possible to regenerate it by immersion for 24 hours in a 1 N solution of hydrochloric acid in 2-propanol. After this treatment the electrode should be washed with distilled water, then with 2-propanol and with methylisobutylketone.

A thick sintered glass or porcelain disc between the saturated potassium chloride solution and the test solution prevents diffusion currents and parasitic potentials.

8. Equivalence point/inflexion point

The equivalence point usually corresponds approximately to the reading 10 on the pH scale, and can be determined graphically by observing the inflexion point on the neutralization curve. Alternatively, it can be calculated as the figure for which the first differential of the variation of pH as a function of the amount of *iso*propanolic potassium hydroxide solution added reaches a maximum, or the value for which the second differential becomes zero.

N.B. It is not possible to determine the inflexion point in the case of crude cottonseed oils rich in gossypol.

In this case the inflexion point may be taken arbitrarily as the pH corresponding to the equivalence point when oleic acid is neutralized by potassium hydroxide in the same solvent as that used for the titration.

Dissolve about 0.282 g of oleic acid in 50 ml of methylisobutylketone (5.4.2). Construct the curve for neutralization of the oleic acid by the potassium hydroxide solution (5.4.4 or 5.4.5) to be used. Read off on the curve the pH of the inflexion point (representing theoretically the addition of 10 ml of a 0.1 N solution of potassium hydroxide).

Apply this figure to the neutralization curve of the cottonseed oil to deduce the amount of potassium hydroxide solution required to "neutralize" the cottonseed oil.

2.202 Determination of the Saponification Value (S.V.)

(Fifth Edition : Method II.D.2)

1. SCOPE

This Standard describes two methods for the determination of the saponification value of animal and vegetable oils and fats.

The indicator method should be preferentially chosen to the potentiometric method when it is applicable.

2. FIELD OF APPLICATION

This Standard is applicable to animal and vegetable oils and fats. It is not applicable to waxes.

3. DEFINITION

The saponification value (S.V.) is the number of mg of potassium hydroxide required to saponify 1 g of fat.

4. INDICATOR METHOD

4.1. Principle

Boiling of the sample under reflux condenser with ethanolic potassium hydroxide solution, and titration of the excess potassium hydroxide with hydrochloric acid in the presence of an indicator.

4.2. Apparatus

4.2.1. 250-ml round-bottomed flasks, or conical flasks, made of alkali-resistant glass.

4.2.2. 1-m reflux tubes, or 30 cm minimum in length reflux condensers, to fit the round-bottomed flasks or conical flasks (4.2.1).

4.2.3. 25-ml volumetric pipette.

4.2.4. 50-ml burette, graduated in 0.1 ml.

4.2.5. Boiling chips.

4.2.6. Heating device (water bath, electric hot-plate,...).

4.3. Reagents

4.3.1. Potassium hydroxide, approximately 0.5 N solution in 95 per cent (V/V) ethanol. Use a solution prepared at least 5 days previously and decanted into a bottle of brown glass, provided with a rubber stopper. The solution should be colourless or straw yellow *(note 1)*.

4.3.2. Hydrochloric acid, 0.5 N aqueous solution, accurately standardized.

4.3.3. Phenolphthalein, 10 g/l solution in 95 per cent (V/V) ethanol.

4.4. Procedure

Prepare the sample according to 2.001, "Preparation of the sample".

Into the round-bottomed or conical flask (4.2.1) weigh to the nearest 0.005 g about 2 g of the prepared sample. With the pipette (4.2.3) add 25 ml of the ethanolic potassium hydroxide solution (4.3.1) and some boiling chips (4.2.5). Fit a reflux tube or reflux condenser (4.2.2). Boil gently with occasional shaking.

After 60 minutes, stop heating. Add to the hot solution 0.5 to 1.0 ml of phenolphthalein (4.3.3) and titrate with the hydrochloric acid solution (4.3.2) until the colour of the indicator changes.

Carry out two determinations with the same prepared sample.

Carry out a blank test in the same way.

4.5. Expression of Results

The saponification value (S.V.) is given by the formula :

$$S.V. = \frac{56.1 \times T \times (V_0 - V_1)}{m}$$

where V_0 is the number of ml of the standardized hydrochloric acid solution (4.3.2) used for the blank test,
V_1 is the number of ml of the standardized hydrochloric acid solution (4.3.2) used for the test with the fat,
T is the exact normality of the standardized hydrochloric acid solution (4.3.2) used,
m is the mass, in g, of the test portion.

Take as the result the arithmetic mean of the two determinations, provided that the requirement of repeatability 4.6 is satisfied.

4.6. Repeatability

The difference between the results of two determinations carried out simultaneously or in rapid succession by the same analyst should not exceed 0.5 per cent of the mean value.

5. POTENTIOMETRIC METHOD

5.1. Foreword

This method should be employed instead of method 4 for the determination of the saponification value of highly coloured oils and fats.

5.2. Principle

Reflux boiling of the sample with an *iso*propanolic potassium hydroxide solution and potentiometric titration of the excess of potassium hydroxide by hydrochloric acid in a non-aqueous medium.

5.3. Apparatus

5.3.1. 100-ml flasks with ground joints, made of alkali-resistant glass.

5.3.2. 1-m reflux tubes, or 30 cm minimum in length reflux condensers, to fit the flasks (5.3.1).

5.3.3. 150-ml tall form beakers.

5.3.4. 1000-ml volumetric flasks.

5.3.5. 25-ml volumetric pipette.

5.3.6. 50-ml burette, graduated in 0.1 ml.

5.3.7. pH meter equipped with glass and calomel electrodes (*note 2*). Contact between the saturated potassium chloride solution and the test solution is made across a sintered glass or porcelain disc at least 0.3 cm thick.

5.3.8. Stirrer, preferably a magnetic stirrer.

5.3.9. Boiling chips.

5.4. Reagents

5.4.1. 2-Propanol (*iso*propanol), analytical reagent quality.

5.4.2. Ethanediol, analytical reagent quality.

5.4.3. Acetic acid, analytical reagent quality.

5.4.4. Sodium carbonate, analytical reagent quality.

5.4.5. Potassium hydroxide, approximately 0.5 N solution in 2-propanol (5.4.1) : dissolve 35 g of potassium hydroxide pellets in 1000 ml of 2-propanol (5.4.1).

5.4.6. Hydrochloric acid, 0.5 N accurately standardized solution in 2-propanol (5.4.1). Pour 42 ml of hydrochloric acid (ρ = 1.18) into a volumetric flask (5.3.4) and make up to the mark with 2-propanol (5.4.1).
Standardization of the solution : e.g. weigh accurately to within 0.0002 g about 0.35 g of sodium carbonate (5.4.4), which has been previously calcined to constant weight at 270-300°C. Transfer to a beaker (5.3.3). Add gradually and cautiously a slight excess of acetic acid (5.4.3) to decompose the sodium carbonate completely. When the evolution of carbon dioxide has ceased, add sufficient distilled water to dissolve the sodium acetate formed and evaporate the solution to dryness on a water bath. Dissolve the residue in 20 ml of 2-propanol (5.4.1). Add 20 ml of ethanediol (5.4.2). Titrate by whichever of the two methods described below is preferred :

(a) Insert the electrodes of the pH meter (5.3.7) and start the stirrer (5.3.8). Titrate with the *iso*propanolic solution of hydrochloric acid (5.4.6) to the equivalent point (*note 3*).

(b) Add 2 drops of a 10 g/l solution of thymol blue in 2-propanol (5.4.1) and titrate with the *iso*propanolic solution of hydrochloric acid (5.4.6) until the colour changes from yellow to pink.

$$\text{Normality} = \frac{1000 \times m_0}{53 \times a_0}$$

where a_0 is the number of ml of the *iso*propanolic solution of hydrochloric acid (5.4.6) used,

m_0 is the mass, in g, of sodium carbonate taken.

5.5. Procedure

Prepare the sample according to 2.001.

Into a ground-necked flask (5.3.1) weigh to within 0.002 g about 2 g of the prepared fat. Add with the pipette (5.3.5) 25 ml of the *iso*propanolic potassium hydroxide solution (5.4.5). Fit the tube or the condenser (5.3.2) and boil gently, with occasional shaking, until the solution become homogeneous.

Stop heating after 60 minutes. In order to avoid solidification of the saponified solution, transfer it quantitatively, while still warm, into a beaker (5.3.3) with the aid of three 10-ml portions of ethanediol (5.4.2). Finish rinsing with 5 ml of 2-propanol (5.4.1).

Insert the electrodes of the pH meter (5.3.7) and start the stirrer (5.3.8). Titrate with the *iso*propanolic hydrochloric acid solution (5.4.6) to the equivalence point of the neutralization curve (*note 3*).

Carry out two determinations on the same prepared sample.

Carry out a blank test in the same way.

5.6. Expression of Results

The saponification value (S.V.) is given by the formula :

$$\text{S.V.} = \frac{56.1 \times T \times (V_0 - V_1)}{m}$$

where V_0 is the number of ml of the standardized *iso*propanolic hydrochloric acid solution (5.4.6) used for the blank test,

V_1 is the number of ml of the standardized *iso*propanolic hydrochloric acid solution (5.4.6) used for the test with the fat,

T is the exact normality of the *iso*propanolic hydrochloric acid solution (5.4.6) used,

m is the mass, in g, of the test portion.

Take as the result the arithmetic mean of the two determinations, provided that the requirement of repeatability 5.7 is satisfied.

5.7. Repeatability

The difference between the results of two determinations carried out simultaneously or in rapid succession by the same analyst should not exceed 0.5 per cent of the mean value.

6. TEST REPORT

The test report should show the method used.

7. NOTES

1. A stable colourless solution of potassium hydroxide can be prepared in the following manner: reflux 1000 ml of ethanol with 8 g of potassium hydroxide and 5 g of aluminium pellets for 1 hour, then distil immediately. Dissolve the required amount of potassium hydroxide in the distillate. Allow the whole to stand for several days and decant the clear supernatant liquid from the deposite potassium carbonate.

   The solution can also be prepared without distillation in the following manner : add 4 ml of aluminium butylate to 1000 ml ethanol and allow the mixture to stand for several days. Decant the supernatant liquid and dissolve therein the necessary amount of potassium hydroxide. The solution is ready for use.

2. It is advisable to store the glass electrode in water or, better still, in a 6/1 (V/V) mixture of ethanediol (5.4.2) and 2-propanol (5.4.1) for 12 hours preceding the titration. Dry it very gently with a piece of filter paper before use. Immediately after the titration, rinse with ethanediol, then with 2-propanol, and finally with distilled water.

   If the electrode does not function satisfactorily, it may be possible to regenerate it by immersion for 24 hours in a 1 N solution of hydrochloric acid in 2-propanol. After this treatment the electrode should be washed with distilled water, then with 2-propanol and with a 6/1 (V/V) mixture of ethanediol and 2-propanol.

   A thick sintered glass or porcelain disc between the saturated potassium chloride solution and the test solution prevents diffusion currents and parasitic potentials.

3. Equivalence point/inflexion point

   The equivalence point usually corresponds approximately to the reading 10 on the pH scale, and can be determined graphically by observing the inflexion point on the neutralization curve. Alternatively, it can be calculated as the figure for which the first differential of the variation of pH as a function of the amount of *iso*propanolic hydrochloric acid solution added reaches a maximum, or the value for which the second differential becomes zero.

2.203. Determination of the Ester Value (E.V.)

(Fifth Edition : Method II.D.3)

1. SCOPE

This Standard describes a method for the characterization of esters in oils and fats.

2. FIELD OF APPLICATION

This Standard is applicable to the animal and vegetable oils and fats. It is not applicable to waxes.

3. DEFINITION

The ester value (E.V.) is the number of mg of potassium hydroxide required to saponify the esters contained in 1 g of the fat.

4. PROCEDURE

For the fat to be analysed, determination of the acid value (A.V.) according to 2.201 and the saponification value (S.V.) according to 2.202.

5. EXPRESSION OF RESULTS

The ester value (E.V.) is given by the formula :

$$E.V. = S.V. - A.V.$$

2.204. Determination of Volatile Acids

(Reichert-Meissl, Polenske and Kirschner Values)

(Fifth Edition : Method II.D.9)

1. SCOPE

This Standard describes the determination of some values able to characterize the low molecular weight fatty acids.

The methods for the determination of volatile acids are empirical, for they determine only a part of these acids. Consequently, the directions must be followed meticulously in order to obtain accurate and reproducible results.

2. FIELD OF APPLICATION

The Standard is applicable to animal and vegetable oils and fats. It is not applicable to waxes.

3. DEFINITIONS

The *Reichert-Meissl value* (soluble volatile acids value) is the number of millilitres of 0.1 N aqueous alkali solution required to neutralize the water-soluble volatile acids isolated from 5 g of the oil or fat under the prescribed conditions.

The *Polenske value* (insoluble volatile acids value) is the number of millilitres of 0.1 N aqueous alkali solution required to neutralize the water-insoluble acids isolated from 5 g of the oil or fat under the prescribed conditions.

The *Kirschner value* is the number of millilitres of 0.1 N aqueous alkali solution required to neutralize the water-soluble volatile acids, which form water-soluble silver salts, isolated from 5 g of the oil or fat under the prescribed conditions.

4. PRINCIPLE

Saponification of the sample in a glycerolic solution, then dilution by water and liberation of the free fatty acids by sulphuric acid. In rigorously specified conditions distillation of the fatty acids which are steam distillable, separation of the water-soluble and water-insoluble steam-distillated fatty acids, and determination of these.

5. APPARATUS

5.1. 25-ml beaker.

5.2. 100-ml graduated cylinder.

5.3. 50-ml pipette.

5.4. Graduated burette.

5.5. Glass apparatus for distillation as shown in the diagram. The apparatus comprises:

Flat-bottomed flask (A) of capacity about 300 ml (diameter of the base of the flask is 45 ± 5 mm).

Stillhead (B) of wall-thickness 1.25 ± 0.25 mm.

Condenser (C), the thickness of the wall of the inner tube shall be 1.0 ± 0.2 mm and that of the exterior walls 1.25 ± 0.25 mm.

Two-mark volumetric flask (D) with graduations of the neck at 100 and 110 ml.

Sheet of asbestos or similar material (E) to support the flask (A) and to protect the sides of the flask from direct flame, diameter 120 mm and thickness 6 mm, with a central hole 65 mm in diameter.

5.6. Heating device (Bunsen burner) such as to allow the distillation to be carried out within the specified time without readjustment.

5.7. Water bath.

5.8. Filter paper, 90 mm diameter.

5.9. Pumice powder; the particles should pass through a sieve with round holes of diameter 2 mm and be retained on a sieve with round hole 1.4 mm diameter.

6. REAGENTS

6.1. Distilled water, freed from carbon dioxide by boiling for 15 minutes.

6.2. Ethanol, 95 per cent (V/V).

6.3. Glycerol, 98 per cent (ρ = 1.26), neutral to phenolphthalein.

6.4. Silver sulphate powder.

6.5. Sodium hydroxide, 50 per cent (m/m) aqueous solution. Before use, decant while protected from carbon dioxide; the solution as used must be limpid.

6.6. Sulphuric acid, approx. N aqueous solution.

6.7. Sodium or potassium hydroxide, 0.1 N aqueous solution, accurately standardized.

6.8. Barium hydroxide, 0.1 N aqueous solution, accurately standardized.

6.9. Phenolphthalein, 10 g/l solution in ethanol (6.2).

7. PREPARATION OF THE SAMPLE

Prepare the sample according to 2.001, "Preparation of the sample".

8. METHODS

8.1. Determination of the Reichert-Meissl Value (Water-Soluble Volatile Acids)

Into the flask (A) of the apparatus (5.5) weigh accurately, to within ± 0.01 g, 5.0 g of the prepared fat. Add 20 g (about 16 ml) of glycerol (6.3.). Add 2 ml of the sodium hydroxide solution (6.5) from the burette (5.4) protected against the entry of carbon dioxide; clean the nozzle before use by discarding the first few drops from the tap.

Heat the flask gently over a naked flame until the fat is melted and then with continuous shaking until the fat, including any droplets adhering to the upper part of the flask, is saponified and the liquid has become perfectly clear; avoid overheating during the process.

Allow to cool to about 90°C. Add 93 ml of boiling distilled water (6.1) from the graduated cylinder (5.2), and mix by shaking. The liquid must remain limpid. If the solution is not clear (incomplete saponification) or if it is darker than a clear yellow (overheated), repeat the saponification on a fresh test portion. Add 0.1 g of the pumice powder (5.9), then 50 ml of the aqueous sulphuric acid (6.6). Connect the flask to the distillation apparatus.

Heat very gently until the liberated fatty acids are completely melted. Then set the flame so that 110 ml of distillate shall be collected within 19 to 21 minutes. The beginning of distillation is to be taken as the moment when the first drop appears at the condenser end of the stillhead (B). Collect the distillate in the graduated flask (D). The temperature of the cooling water leaving the condenser should be between 15 and 20°C.

When the distillate exactly reaches the 110-ml mark on the flask, remove the flame and quickly replace the flask by the beaker (5.1). Stopper the graduated flask and place it in a water bath (5.7) maintained at 15°C, so that the 110-ml graduation is 1 cm below the water level in the bath; keep it there for 10 minutes.

Remove the graduated flask from the water bath, dry the outside and mix the contents by inverting the flask 4 or 5 times without shaking; avoid wetting the stopper with the insoluble acids. Filter the liquid through a dry, plain filter paper (5.8), fitting snugly into a funnel. The filtrate should be clear (*note 1*). Take a measured 100 ml of the filtrate and add 5 drops of the phenolphthalein solution (6.9).

(a) If the Kirschner value is to be determined, titrate with the aqueous barium hydroxide solution (6.8); in this case the titration flask must be dry before use. Record the precise volume of the barium hydroxide solution used for the titration; if this is less than 5 ml, add 5 ml of distilled water (*note 2*). Close the flask with a cork and retain the solution for the determination of the Kirschner value as described in 8.3.

(b) If the determination of the Kirschner value is not required, aqueous sodium or potassium hydroxide solution (6.7) may be used for the titration if desired.

Carry out a blank test without fat, but using the same amounts of reagents and the same procedure.

8.2. Determination of the Polenske Value (Water-Insoluble Volatile Acids)

After titrating the soluble volatile acids, detach the stillhead (B) and rinse the condenser (C) with 15 ml of distilled water (6.1) at 15°C. Collect the washing in the beaker (5.1). Rinse the 100/110-ml graduated flask with this liquid, pour it onto the filter and allow all of it to pass through.

Repeat this operation twice more, using 15 ml of distillated water each time. The size of the filter is such that each 15-ml portion can be poured straight into it and will fill it to the brim.

The last wash is collected separately and should require no more than one drop of alkali solution (6.7) for neutralization. Reject the aqueous washings.

Place the funnel in the neck of a clean, dry flask. Dissolve the water-insoluble fatty acids by washing the condenser with 15 ml of ethanol (6.2). Collect the wash in the beaker (5.1). Then use the liquid to rinse the volumetric flask (D) and pour it in one lot onto the filter. Wash twice more in the same way.

Titrate the total ethanolic washings (45 ml in all), after the addition of 3 drops of phenolphthalein solution (6.9), with an aqueous solution of barium (6.8), sodium or potassium hydroxide (6.7).

8.3. Determination of the Kirschner Value

Add 0.5 g of the powdered silver sulphate (6.4) to the solution, neutralized with the aqueous barium hydroxide (6.8) as described in 8.1 (determination of Reichert-Meissl value).

Allow the flask to stand in the dark for 1 hour with occasional shaking.

Filter the contents away from the light through a dry filter. Transfer 100 ml of the filtrate into a dry 300-ml flat-bottomed flask of the same type as that used for the determination of the volatile acids, i.e. flask (A) (*note 2*). Add 35 ml of cold distilled water (6.1) recently boiled for 15 minutes, 10 ml of the aqueous sulphuric acid solution (6.6) and 0.1 g of pumice powder (5.9) (*note 3*).

Attach the flask to the distillation apparatus (5.5); start heating and distil, as before, so that 110 ml of distillate are collected in 19-21 minutes. The beginning of distillation is taken as the moment when the first drop is formed at the lower (condenser) tip of the stillhead (B). Collect the distillate in the graduated flask (D). When the distillate reaches the 110-ml mark, remove the flame and quickly replace the flask by the beaker (5.1).

Stopper the flask (D) and mix the distillate by inverting the flask 4 or 5 times without shaking. Filter the liquid through a dry plain filter (5.8) fitting snugly in a funnel. The filtrate should be clear. Measure 100 ml, add 5 drops of phenolphthalein solution (6.9) and titrate with the aqueous barium hydroxide solution (6.8).

Carry out a blank test on the solution from the blank test described in 8.1, and neutralized with the aqueous barium hydroxide solution (6.8).

9. EXPRESSION OF RESULTS

9.1. Reichert-Meissl Value (Water-Soluble Volatile Acids) : RM.V.

The Reichert-Meissl value (RM.V.) is given by the formula :

$$\text{RM.V.} = 11 \times T \times (a - b) \qquad (\textit{note 4})$$

where a is the number of ml of the standardized alkali solution* used for the test portion,

b is the number of ml of the standardized alkali solution* used for the blank test,

T is the exact normality of the alkali solution* used.

* Aqueous solution of barium (6.8), potassium or sodium (6.7) hydroxide.

9.2. Polenske Value (Water-Insoluble Volatile Acids) : P.V.

The Polenske value (P.V.) is given by the formula :

$$P.V. = 10 \times T \times c \qquad \textit{(note 4)}$$

where c is the number of ml of the standardized alkali solution* used for the test portion,

T is the exact normality of the alkali solution* used.

* Aqueous solution of barium (6.8), potassium or sodium (6.7) hydroxide.

9.3. Kirschner Value : K.V.

The Kirschner value (K.V.) is given by the formula :

$$K.V. = 0.121 \times T \times (d - e) \times (100 + a)$$

where a is the number of ml of the standardized aqueous barium hydroxide solution used for the determination of the Reichert-Meissl value,

d is the number of ml of the standardized aqueous barium hydroxide solution (6.8) used for the test portion,

e is the number of ml of the standardized aqueous barium hydroxide solution (6.8) used for the blank test,

T is the exact normality of the aqueous barium hydroxide solution (6.8) used.

10. NOTES

1. After filtration, the filtrate should be free from insoluble fatty acids. It is, however, to be noted that this cannot easily be achieved when one is operating on coconut oil. In this case, liquid insoluble fatty acids may pass through the filter. This may be avoided by using a wetted filter paper, or by collecting the filtrate in a separating funnel and, after settling, removing the lower aqueous layer, the insoluble acids will float at the surface. Add these to the bulk of the insoluble acids.

2. If the Reichert-Meissl value is low, it is possible that, after titration with the barium hydroxide solution (6.8), addition of the powdered silver sulphate (6.4) and filtration, the volume of the filtrate may not be enough to provide 100 ml for the distillation flask.

   In this case, 5 ml of distilled water (6.1) may be added to the filtered solution and a factor (105 + a) substituted for (100 + a) in the equation 9.3.

3. In the determination of the Kirschner value, the pumice may be replaced by a 5-mm coil of aluminium wire, 30 cm long and 1 mm dia., as is usual in British practice.

4. The Polenske value, and to a less extent the Reichert-Meissl value, may come out too low if the determination is carried out at low barometric pressure such as may occur at high altitudes.

The following formulae may be used to correct the observed values found at a barometric pressure of h mm Hg to values appropriate to normal pressure :

Corrected Reichert-Meissl value :

$$\frac{(\text{Observed value} - 10) \times \log 760}{\log h} + 10.$$

Corrected Polenske value :

$$\text{Observed value} \times \frac{760 - 45}{h - 45}.$$

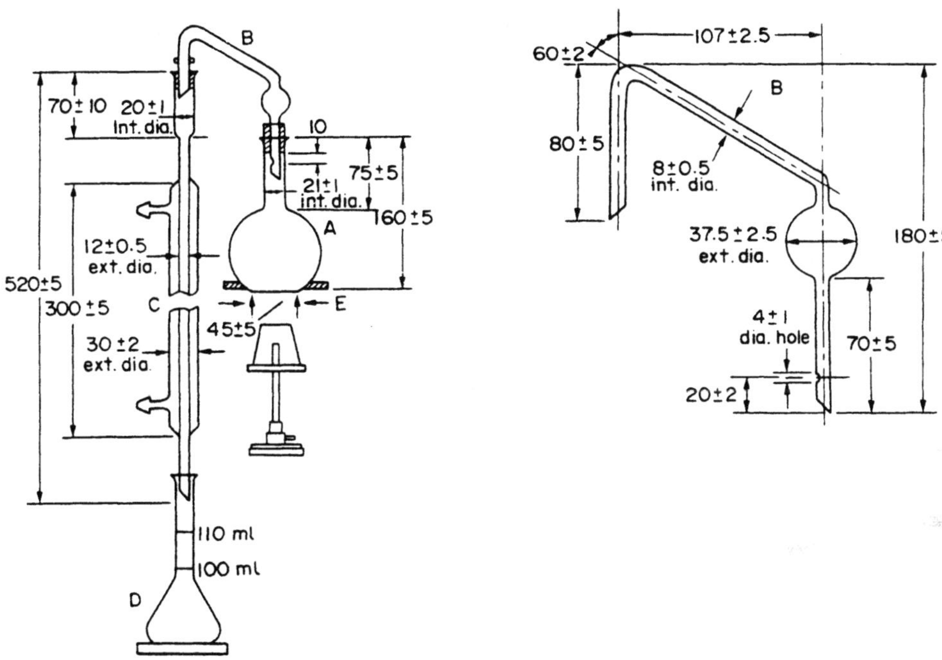

Distillation apparatus (dimensions in millimetres).

2.205. Determination of the Iodine Value (I.V.)

(Fifth Edition : Method II.D.7)

1. SCOPE AND FOREWORD

This Standard describes some methods for the determination of the iodine value of oils and fats.

Three methods are given, namely those of Wijs, Hanus and Hübl.

The methods of Wijs and Hanus are recommended for industrial and commercial analysis.

The Hübl method is still official in some countries (except for commercial transactions).

The statement of results must mention the method used, as the different methods give different figures for the same oil.

2. FIELD OF APPLICATION

This Standard is applicable to animal and vegetable oils and fats, and to waxes.

3. DEFINITION

The iodine value (I.V.) of a fat is the number of grammes of halogen absorbed by 100 grammes of the fat, and expressed as the weight of iodine.

4. THE WIJS METHOD

4.1. Principle

Addition to the test portion of an iodine monochloride solution in an acetic and carbon tetrachloride mixture. After a standard time of reaction, determination of the excess halogen by addition of potassium iodide aqueous solution and titration of the liberated iodine with a standardized sodium thiosulphate solution.

4.2. Apparatus

The apparatus must be scrupulously clean and perfectly dry.

4.2.1. 2 to 4-ml glass weighing scoops.

4.2.2. 250-ml wide-necked glass bottles, with ground-glass stoppers.

4.2.3. 50-ml burette, graduated in 0.1 ml.

4.3. Reagents

4.3.1. Glacial acetic acid, free from ethanol.

4.3.2. Carbon tetrachloride, analytical reagent quality.
The absence of oxidizable matter in both the above reagents is checked by shaking 10 ml of the reagent with 1 ml of saturated aqueous potassium dichromate solution and 2 ml of concentrated sulphuric acid. No green coloration should appear.

4.3.3. Iodine trichloride (ICl_3), or, failing this, iodine monochloride (ICl).

4.3.4. Iodine, pure, re-sublimed.

4.3.5. Potassium iodide, 100 g/l aqueous solution.
This must contain no iodine or iodate.

4.3.6. Hydrochloric acid 1/1 (V/V) aqueous solution, if the Wijs reagent is made from iodine monochloride.

4.3.7. Sodium thiosulphate, 0.1 N accurately standardized aqueous solution.

4.3.8. Potassium iodate, 0.4 N accurately standardized aqueous solution, if the Wijs reagent is made from iodine monochloride.

4.3.9. Starch solution, 10 g/l aqueous dispersion, recently prepared from natural soluble starch.

4.4. Preparation of the Wijs Reagent

4.4.1. Preparation of the Wijs reagent with iodine trichloride

Weigh 9 g of iodine trichloride (4.3.3) into a brown glass bottle (1500 ml capacity). Dissolve in 1 l of a mixture composed of 700 ml of acetic acid (4.3.1) and 300 ml of carbon tetrachloride (4.3.2).

Determine the halogen content by the following method :

Take 5 ml of the solution and add 5 ml of the aqueous solution of potassium iodide (4.3.5) and 30 ml of water. Titrate with the sodium thiosulphate solution (4.3.7) in the presence of a few drops of starch solution (4.3.9) as indicator.

Add 10 g of powdered iodine (4.3.4) to the bulk of the reagent and dissolve by shaking.

Determine the halogen content as above. The titre should now equal one and half times that of the first determination. If this is not the case, add a small quantity of iodine until the content slightly exceeds the limit of one and half times. It is important that no trace whatever of iodine trichloride should remain as it would cause secondary reactions.

Let the solution stand, then decant the clear liquid into a brown glass bottle. If stored in a well-stoppered bottle away from light, the solution can be used for several months.

4.4.2. Preparation of the Wijs reagent with iodine monochloride

In the lack of iodine trichloride, the Wijs reagent may be prepared with iodine monochloride as follows :

Dissolve 19 g of iodine monochloride (4.3.3) into a brown glass bottle (1500 ml capacity) in 1 l of a mixture of 700 ml acetic acid (4.3.1) and 300 ml carbon tetrachloride (4.3.2). After the addition of a few milligrammes of pure iodine (4.3.4), determine the content of free halogen as described in 4.4.1. Dilute, if necessary, with the specified mixture of solvents until 5 ml of the solution are equivalent to about 10 ml of the sodium thiosulphate solution (4.3.7).

The exact standardization of this reagent must be tested as follows:

(a) Into a wide-necked flask (4.2.2) put 50 ml of hydrochloric acid solution (4.3.6) and 50 ml of carbon tetrachloride (4.3.2). Add exactly 25 ml of Wijs reagent prepared with iodine monochloride and shake. Titrate the free iodine present in the reddish-violet layer of carbon tetrachloride with the potassium iodate solution (4.3.8), shaking vigorously until the layer becomes colourless.

N.B. If the carbon tetrachloride layer is colourless, this indicates that the Wijs reagent does not contain free iodine and consequently the ratio between the iodine and chlorine is less than 1/1; in this case it is necessary to add to 25 ml of the prepared Wijs solution a quantity of pure iodine so that the carbon tetrachloride layer develops a reddish-violet coloration. Next, calculate the quantity of iodine necessary for the total amount of Wijs solution and dissolve this in the solution in question. Proceed once again as described above and titrate the free iodine in the reddish-violet layer of carbon tetrachloride with the potassium iodate solution (4.3.8).

(b) Into a second wide-necked flask (4.2.2) introduce exactly 25 ml of Wijs solution containing a sufficient amount of free iodine. Add 15 ml of the aqueous potassium iodide solution (4.3.5) and about 150 ml of distilled water.

Shake and titrate the liberated iodine with the sodium thiosulphate solution (5.3.4) in the presence of a few drops of starch solution (4.3.9) as indicator; at the end of the titration shake vigorously.

Calculation :

$$\frac{\text{iodine}}{\text{chlorine}} = \frac{V_1T_1 + V_2T_2}{V_1T_1 - V_2T_2}$$

where V_1 is the number of ml of the standardized sodium thiosulphate solution (4.3.7) used to determine total halogen,

V_2 is the number of ml of the standardized potassium iodate solution (4.3.8) used to determine free iodine,

T_1 is the exact normality of the sodium thiosulphate solution (4.3.7) used,

T_2 is the exact normality of the potassium iodate solution (4.3.8) used.

Comments

As the point at issue is to determine only whether the Wijs solution really contains a slight excess of iodine, one can stop at the titration described under point (a) of the procedure. At the moment that the carbon tetrachloride layer becomes coloured, either directly after adding 25 ml of the Wijs solution to the mixture of hydrochloric acid solution and carbon tetrachloride, or after having dissolved a sufficient amount of pure iodine in the Wijs solution, it is evident that the ratio to be determined is more than 1.

4.5. Procedure

The weight of fat to be taken varies according to its expected iodine value in the following way :

Iodine value expected	Weight to be taken for test (g)
< 5	3.00
5 to 20	1.00
21 to 50	0.40
51 to 100	0.20
101 to 150	0.13
151 to 200	0.10

In the glass weighing scoop (4.2.1) weigh the appropriate quantity of fat to within 0.001 g. Place it in the ground-necked bottle (4.2.2). Add 15 ml of carbon tetrachloride (4.3.2) to dissolve the fat. Add exactly 25 ml of the reagent, insert the stopper, shake gently and place the bottle in the dark.

For fats having an iodine value below 150, leave the bottle in the dark for 1 hour; for those with an iodine value above 150 and for polymerized or oxidized fats, leave for 2 hours.

At the end of this time, add 20 ml of the potassium iodide solution (4.3.5) and 150 ml of water. Titrate with the sodium thiosulphate solution (4.3.7), using starch solution (4.3.9) as indicator, continuing the titration until the blue colour just disappears after very vigorous shaking.

Carry out a blank test simultaneously without the fat under the same conditions.

4.6. Expression of Results

The iodine value (I.V.) is given by the formula :

$$\text{I.V.} = \frac{12.69 \times T \times (V_3 - V_4)}{m}$$

where V_3 is the number of ml of the standardized sodium thiosulphate solution (4.3.7) used for the blank test,

V_4 is the number of ml of the standardized sodium thiosulphate solution (4.3.7) used for the test portion.

T is the exact normality of the sodium thiosulphate solution (4.3.7) used,

m is the mass, in g, of the test portion.

5. THE HANUS METHOD

5.1. Principle

Addition to the test portion of an iodine monobromide solution in an acetic acid and carbon tetrachloride mixture. After a standardized time of reaction, determination of the excess halogen by addition of potassium iodide aqueous solution and titration of the liberated iodine with a standardized sodium thiosulphate solution.

5.2. Apparatus

See 4.2.

5.3. Reagents

5.3.1. Glacial acetic acid, free from ethanol.

5.3.2. Carbon tetrachloride, analytical reagent quality.
The two reagents must be free from oxidizable matter and are tested as in 4.3.

5.3.3. Iodine monobromide (IBr), pure.

5.3.4. Potassium iodide, 100 g/l aqueous solution.
This must contain no free iodine or iodate.

5.3.5. Sodium thiosulphate, 0.1 N accurately standardized aqueous solution.

5.3.6. Starch solution, 10 g/l aqueous dispersion, recently prepared from natural soluble starch.

5.4. Preparation of the Hanus Reagent

Into a brown glass bottle weigh 10 g of iodine monobromide (5.3.3) and dissolve it in 500 ml of acetic acid (5.3.1).

5.5. Procedure

The weight of fat to be taken, in g, is about 25/I.V. (expected iodine value).

Into a glass scoop (4.2.1) weigh the appropriate quantity accurately to within 0.001 g. Insert the scoop with contents into a ground-necked bottle (4.2.2). Add 10 ml of carbon tetrachloride (5.3.2) and dissolve.

Then add exactly 25 ml of the reagent and insert the stopper. Shake gently, and place in the dark for 1 hour.

Add 20 ml of the potassium iodide solution (5.3.4) and 100 ml of distilled water. Titrate with the aqueous thiosulphate solution (5.3.5), using starch solution (5.3.6) as indicator, continuing the titration until the blue colour just disappears after very vigorous shaking.

5.6. Expression of Results

See 4.6.

6. THE HÜBL METHOD

6.1. Principle

To a carbon tetrachloride solution of the test portion, addition of an excess of an ethanolic solution of iodine and mercuric chloride. After a standardized time of reaction, determination of the excess halogen by addition of potassium iodide aqueous solution and titration of the liberated iodine with a standardized sodium thiosulphate solution.

6.2. Apparatus

See 4.2.

6.3. Reagents

6.3.1. Ethanol, 95 per cent (V/V).

6.3.2. Carbon tetrachloride, free from oxidizable matter (see 4.3).

6.3.3. Iodine, pure, re-sublimed.

6.3.4. Mercuric chloride, analytical reagent quality.

6.3.5. Potassium iodide, 300 g/l aqueous solution.
This must contain no free iodine or iodate.

6.3.6. Sodium thiosulphate, 0.1 N accurately standardized aqueous solution.

6.3.7. Starch solution, 10 g/l aqueous dispersion, recently prepared from natural soluble starch.

6.4. Preparation of the Hübl Reagent

Into a brown glass bottle dissolve 25 g iodine (6.3.3) in 500 ml of ethanol (6.3.1).

Into another bottle weigh 20 g mercuric chloride (6.3.4) and dissolve in 500 ml of ethanol (6.3.1). Filter this solution, if necessary.

24 hours before it is required for use, mix equal volumes of these two solutions. This mixture does not keep for more than 48 hours.

6.5. Procedure

The weight of fat to be taken varies according to its expected iodine value according to the table given in 4.5.

In a small glass scoop (4.2.1) weigh the appropriate quantity of fat accurately to within 0.001 g. Insert into a 500-ml ground-necked bottle.

Dissolve in 10 ml of carbon tetrachloride (6.3.2). Add exactly 25 ml of the reagent. Insert the stopper, shake gently and place the bottle in the dark.

For oils having an iodine value below 150, leave for 12 hours; for those with iodine value above 150 and for polymerized or oxidized oils, leave for 24 hours.

Then add 20 ml of the potassium iodide solution (6.3.5). If a mercuric iodide precipitate appears, add more potassium iodide solution until the precipitate redissolves. Then add 300 ml of distilled water.

Titrate with the aqueous sodium thiosulphate solution (6.3.6) in the presence of starch solution (6.3.7) as indicator, continuing the titration until the blue colour just disappears after very vigorous shaking.

Carry out a blank test without the fat under the same conditions.

6.6. Expression of Results

See 4.6.

7. REPEATABILITY

The difference between the results of two determinations carried out simultaneously or in rapid succession by the same analyst should not exceed 0.4 to 1 unit according to the expected value.

2.206. Determination of Di- and Tri-Unsaturated Fatty Acids by Ultraviolet Spectrophotometry

(Fifth Edition : Method II.D.17)

1. SCOPE

This Standard describes a method for the determination of di- and tri-unsaturated fatty acids (more particularly linoleic and linolenic acids) in animal and vegetable oils and fats.

2. FIELD OF APPLICATION

This Standard is applicable to the determination of unsaturated fatty acids containing two or three double bonds, such as linoleic and linolenic acids, in animal and vegetable oils and fats which contain only, or mainly, natural unsaturated fatty acids, i.e. the *cis*-isomers of poly-unsaturated fatty acids with non-conjugated double bonds. The sample must not contain more than traces of conjugated acids and of constituents (e.g. pigments) whose absorption might undergo considerable change during the alkali-isomerization.

The method, therefore, is *not* applicable *to tung oils and similarly conjugated oils, nor to oxidized oils,* and it can be applied *only with special precautions* to modified drying oils, hydrogenated oils or other oil-containing *trans*-isomers of the unsaturated fatty acids, or to fish oils or similar fats containing appreciable amounts of unsaturated fatty acids with more than three double bonds.

The proposed method is an empirical one. Consequently, in order to obtain reproducible results, strict adherence to the specified operating conditions is essential, particularly with regard to the alkali-isomerization. Constancy of the temperature of the oil bath, the time of heating, the manipulations to obtain a homogeneous mixture, the flow-rate of the nitrogen used, and the concentration of the alkali solution are all very important factors.

3. PRINCIPLE

Measurement of the ultraviolet absorption of solutions of sample before and after alkali-isomerization. This isomerization, by producing conjugation of the double bonds, results in the appearance of characteristic absorption bands in the 200- to 300-nm region.

Calculation from the measured absorption values of the contents of conjugated dienes and trienes originally present and of linoleic and linolenic acids, because an acid with two conjugated double bonds shows an absorption band with a peak at about 230 nm, and with three conjugated double bonds a peak at 268 nm.

4. APPARATUS

4.1. Approx. 1-ml beakers.

4.2. 100-ml volumetric flasks.

4.3. Isomerization tubes : see diagram.

4.4. Apparatus for purifying nitrogen. Any apparatus which satisfactorily eliminates oxygen may be used.

4.5. Oil bath, with stirrer, which can be maintained at a constant temperature of 180 ± 0.5°C.

4.6. UV Spectrophotometer.

4.7. Quartz absorption cells, 1.00 or 4.00 cm (± 0.01 cm) in length. The two cells of each pair must be identical.

5. REAGENTS

5.1. *n*-Hexane, or, failing this, 2,2,4-trimethylpentane (*iso* octane), optically clear.

5.2. Ethanol, 95 per cent (V/V) optically clear.

5.3. Ethanediol, analytical quality reagent.

5.4. Potassium hydroxide, 6.5 - 6.6 per cent (m/m) solution in ethanediol (5.3) : weigh 750 g of ethanediol (5.3) into a 1 l flask of alkali-resistant glass. Throughout the preparation of the reagent, pass a current of nitrogen through the liquid by means of a glass tube reaching to the bottom of the flask, in order to exclude all air and to agitate the liquid gently. Place the flask in an oil bath at 100-150°C. Raise the bath temperature to 190°C and hold it at this temperature for 10 minutes. Allow to cool down to 150°C and cautiously add to the ethanediol 60 g of potassium hydroxide pellets, analytical reagent grade (commercial KOH contains up to 15 per cent (m/m) water), in small portions, 3 to 5 pellets at a time, while swirling the flask to dissolve them completely. Raise the bath temperature to 190°C again and hold at this temperature for 10 minutes. Remove the flask from the oil bath and allow the potassium hydroxide ethanediol solution to cool to room temperature. Stopper the flask and store the reagent in a refrigerator at about 5°C under nitrogen. Before use, check the concentration of potassium hydroxide by titration; the solution should contain 6.5 to 6.6 per cent (m/m) of potassium hydroxide.

6. PREPARATION OF THE SAMPLE

Prepare the sample according to 2.001, "Preparation of the sample".

7. PROCEDURE

7.1. Determination of Initially Conjugated Di- and Tri-unsaturated Fatty Acids

Weigh accurately (within 1 per cent) in a beaker (4.1) sufficient of the sample, so that, after it has been dissolved in ethanol (5.2) or hexane (5.1), and the solution transferred quantitatively into a volumetric flask (4.2) and diluted to 100 ml, it will show an absorbance in the UV of between 0.2 and 0.8 when measured in 1.00 cm quartz cell (4.7). The sample must be completely dissolved and the solution perfectly clear. Measure the absorbance of the solution using a UV spectrophotometer (4.6) at the following wavelengths, using the chosen solvent in the comparison cell :

to determine dienes: 233 nm,

to determine trienes: 262, 268 and 274 nm.

7.2 Determination of Polyunsaturated Fatty Acids (Dienes and Trienes) Which Can Be Conjugated by Isomerization

For each determination use four isomerization tubes (4.3) : two tubes for the sample and two tubes for the blank.

Weigh into each of the four tubes 11.0 ± 0.1 g of the ethanediol solution of potassium hydroxide (5.4) and pass nitrogen through the tubes for about 1 minute to remove air. Then, immerse them in the oil bath (4.5) at 180 + 0.5°C without interrupting the nitrogen stream.

Weigh accurately about 100 mg of the sample to the nearest 1 mg into each of two beakers (4.1). Maintain the ethanediol solution of potassium hydroxide in the isomerization tubes under nitrogen for 20 minutes in the oil bath at 180 ± 0.5°C; drop the beakers containing the sample, and two empty beakers for the blank experiments, into the isomerization tubes; remove the tubes, make sure that the beakers are at the bottom of them and swirl vigorously for several seconds to mix the contents. Replace the tubes in the oil bath and leave for 1 minute. Take the tubes out and swirl again. Replace the tubes in the bath. (If the solution is not clear, repeat the heating and swirling procedure two or three times at 1-minute intervals.) Leave the isomerization tubes in the oil bath for exactly 25 minutes after introducing the beakers.

Remove all the tubes from the bath and put them into a beaker filled with oil at room temperature in order to chill them quickly. After they have cooled to room temperature, wipe the tubes and quantitatively transfer the contents of each into a volumetric flask (4.2), using ethanol (5.2) to rinse the tube, and dilute to volume. Measure the absorbance of the sample solutions in a 1.00-cm cell (4.7) with the spectrophotometer (4.6) at the specified wavelengths, using one of the blank solutions in the reference cell. Further, compare the two blanks, one against the other; if the difference between the absorbance of the two blanks is more than 0.005 repeat the whole determination. If the absorbance of the sample is less than 0.2, use the 4.00-cm cell. If the absorbance is more than 0.8, dilute both samples and blank similarly.

8. EXPRESSION OF RESULTS

8.1. General

For each analysis, the requisite spectrophotometric readings and calculations depend upon the nature of the sample. Two general rules are to be observed :

(a) The corrections for background absorption are only required when no more than traces of di- and tri-unsaturated fatty acids occur in the oils and fats. When larger quantities of these acids are present, background corrections are not required, because their use may lead to erroneous results. It is suggested that, when the specific absorption coefficient of a polyunsaturated fat after isomerization is greater than 1.0, no background correction is necessary.

(b) The calculation gives the amount of linoleic and linolenic acids in the oil or fat. To calculate the content of these acids as a proportion of the fatty acids of the oil or fat, the figure must be multiplied by

$$\frac{100}{\text{percentage content (m/m) of fatty acids}}.$$

8.2. Percentage Content (m/m) of Initially Conjugated Acids

Calculate the specific absorption coefficient (a) for the wavelengths 233, 262, 268 and 274 nm by the formula :

$$a = \frac{A}{b \times m}$$

where A is the observed absorbance,

b is the exact cell length, in cm,

m is the mass, in g, of sample per litre of the final solution used for measurement.

(a) The specific absorption coefficient at 233 nm (a_2) corrected for absorption due to acid or ester groups is given by the formula :

$$a_2 = a_{233} - a_o$$

where a_o is 0.07 for esters, and 0.03 for soaps and fatty acids.

(b) The specific absorption coefficient at 268 nm (a_3) corrected for background absorption is given by the formula :

$$a_3 = 2.8\left[a_{268} - \tfrac{1}{2}\,(a_{262} + a_{274})\right].$$

(c) The percentage content (m/m) of conjugated dienes (C_2) is given by the formula :

$$C_2 = 0.91a_2.$$

The percentage content (m/m) of conjugated trienes (C_3) is given by the formula :

$$C_3 = 0.47a_3.$$

8.3. Percentage Content (m/m) of Linoleic and Linolenic Acids

Calculate the specific absorption coefficient (a') for the wavelengths 233, 262, 268 and 274 nm by the formula :

$$a' = \frac{A'}{b' \times m'}$$

where A' is the observed absorbance after isomerization,

b' is the exact cell length, in cm,

m' is the mass, in g, of sample per litre of the final solution used for measurement.

(a) The specific absorption coefficient at 233 nm (a'_2) corrected for the conjugated dienes originally present and for acid groups is given by the formula :

$$a'_2 = a'_{233} - a_2 - 0.03.$$

(b) The specific absorption coefficient at 268 nm (a'_3) corrected for background absorption and for conjugated trienes originally present is given by the formula :

$$a'_3 = 4.03 \left[a'_{268} - \tfrac{1}{2}(a'_{262} + a'_{274})\right] - a_3.$$

(c) The percentages content (m/m) of linoleic and linolenic acids are given by the formulae:

without background correction : linoleic acid = $1.086a'_2 - 1.2(a'_{268} - a_{268})$,

linolenic acid = $1.980(a'_{268} - a_{268})$.

with background correction : linoleic acid = $1.086a'_2 - 1.2a'_3$,

linolenic acid = $1.980a'_3$.

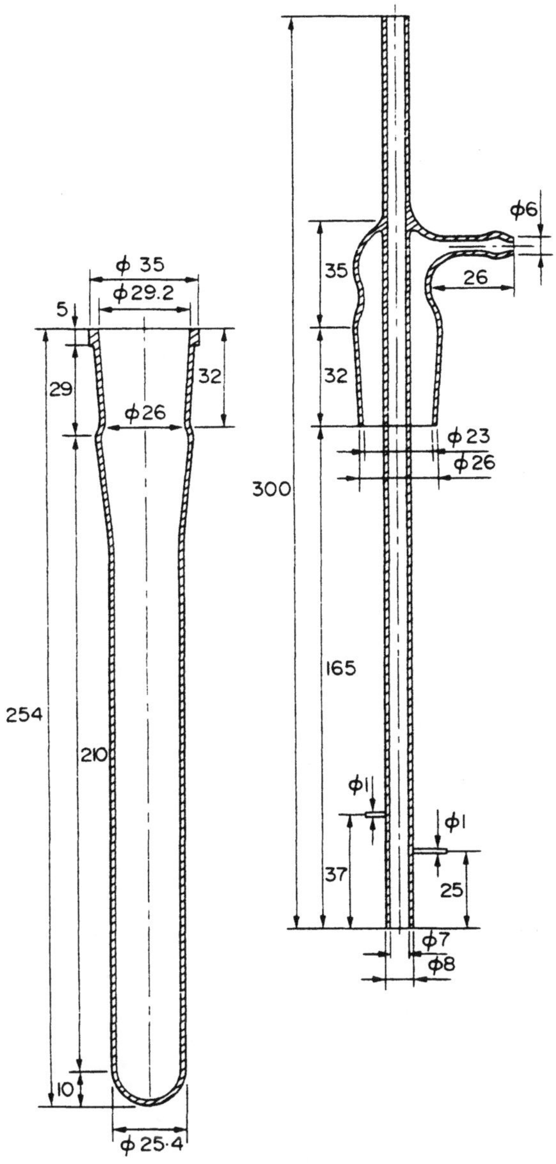

Isomerization tube (dimensions in millimetres).

2.207. Determination of Content of Isolated *trans*-Unsaturated Compounds by Infrared Spectrophotometry

(Fifth Edition : Method II.D.22)

1. SCOPE

This Standard describes a procedure for the determination of the content of isolated (i.e. non-conjugated) *trans*-unsaturated compounds by infrared spectrophotometry.

2. FIELD OF APPLICATION

This Standard is applicable to natural, refined or hydrogenated animal and vegetable oils and fats, fatty acids and fatty acid methyl esters, in which the *trans*-unsaturated compounds are present in more than 1 per cent.

The method yields spectra which can less easily be interpreted, and hence afford less accurate results, in the presence of :

- compounds with functional group absorbing in the neighbourhood of the C — H deformation vibration at 970 cm^{-1} (10.3 nm) or affecting it, e.g. ricinoleic acid in castor oil;
- compounds with conjugated unsaturated double bonds, such as tung oil.

3. PRINCIPLE

Conversion of the oils, fats or fatty acids into fatty acid methyl esters, and measurement of the absorption of the fatty acid methyl esters at around 970 cm^{-1} (10.3 nm), absorption being caused by a deformation of the C — H bonds adjacent to the isolated *trans* double bond.

The content of isolated *trans*-unsaturated components is expressed as percentage of methyl elaidate in the methyl esters.

4. DEFINITION

The content of isolated *trans*-unsaturated components in oils, fats, fatty acids or fatty acid methyl esters is given by that percentage of methyl elaidate present in a mixture of methyl elaidate and methyl stearate for which the specific absorption at 970 cm^{-1} (10.3 nm), as determined by the method prescribed below, is identical with that of the fatty acid methyl esters, either as presented, or as prepared from the oils, fats or fatty acids.

5. APPARATUS

5.1. 10-ml measuring flasks, with ground-glass stopper.

5.2. Recording, double-beam infrared spectrophotometer, suitable for quantitative measurements within a range 1100 and 910 cm^{-1} (9 and 11 nm), with an accuracy of 1 cm^{-1} (0.01 nm).

5.3. 1-mm infrared cells, with windows (e.g. sodium chloride or potassium bromide) which do not absorb in the range of measurement.

The matching pairs of cells required for the calibration solutions or the test solution and for the reference solution should have the same path-length to within 1 per cent.

6. REAGENTS

6.1. Carbon disulphide, anhydrous, analytical reagent quality.
Note : carbon disulphide vapours are poisonous and flammable. Work should be carried out in a well-ventilated room, preferably in a fume cupboard.

6.2. *Standard substance :* methyl elaidate (methyl ester of 9, 10-*trans*-octadecenoic acid), at least 99 per cent pure. Failing this, a mixture with a content of at least 65 per cent isolated *trans*-unsaturated compounds, determined by means of the standard substance, can be used.

6.3. Reference substance : methyl stearate (methyl ester of octadecanoic acid), free from *trans*-unsaturated compounds.

7. PROCEDURE

7.1. Calibration Curve

7.1.1. Preparation of the calibration solutions and the reference solution

Into a series of 10-ml measuring flasks (5.1) weigh, at the nearest 0.1 mg, respectively 0 (reference solution), 25, 50, 100 and 150 mg of the standard substance (6.2), and make up each weighting to 200 mg with methyl stearate (6.3). Dissolve each of these mixtures in carbon disulphide (6.1), and make up to the mark at 20°C.

7.1.2. Measurement of the calibration solutions

N.B. The determinations should be made with the solutions maintained at 20°C.

Rinse a clean cell (5.3) with the calibration solution containing 25 mg of methyl elaidate, and fill it with this solution. If necessary, use a syringe for filling to avoid any formation of air bubbles.

In the same way, fill the second cell (5.3) with the reference solution (containing only methyl stearate).

Place the cell containing the reference solution in one beam of IR spectrophotometer (5.2), and the cell containing the calibration solution in the other beam.

Plot the transmission (or absorption) curve between 1050 and 900 cm^{-1} (9.5-11 nm) at optimal instrumental settings.

Similarly, plot the transmission (or absorption) curves for the other calibration solutions in the order of increasing concentrations.

7.1.3. Construction of the calibration curve

Each of the four transmission (or absorption) curves of the four calibration solutions shows a minimum (or a maximum) at around 970 cm^{-1} (10.3 nm). Draw a base-line for each curve; this is the straight line between the absorption minima at *ca.* 1000 and 925 cm^{-1} (10.0 and 10.8 nm).

For a transmission curve measure the distances *QR* and *PR* (see figure) at the transmission minimum, and calculate the absorption A_i of the solution by the formula :

$$A_i = \log \frac{PR}{QR}.$$

For an absorption curve read the absorptions A_Q at Q, and A_P at P, and calculate the absorption A_i of the solution by the formula :

$$A_i = A_Q - A_P.$$

Plot the graph for the four calibration solutions, taking the amounts of the standard substance in the solutions as abscissae and the corresponding values for the absorption A_i as ordinates. Draw the best straight line through the four points plotted.

For each series of determinations, carry out a control, using one of the calibration solutions, and verify that the measured value for the absorbance lies on the calibration curve. If it does not, construct a fresh calibration curve. Do this, also, if the instrumental settings have been changed, or when parts of the apparatus, or the cells, have been replaced.

7.2. Preparation of the Solution to Be Assayed

Prepare 250-500 mg of methyl esters from the sample of oil, fat or fatty acids according to 2.301, "Preparation of fatty acid methyl esters".

If necessary, melt the methyl esters by warming. Weigh, to the nearest 0.1 mg, 200 mg of the well-mixed preparation into a measuring flask (5.1). Dissolve this test portion in carbon disulphide (6.1) and make up to the mark at 20°C.

If the transmission at 970 cm^{-1} (10.3 nm) of this solution is found to be less than 10 per cent, or if enough sample is not available, weigh a smaller test portion and add sufficient methyl stearate (6.3) to make up the total amount to 200 mg.

7.3. Measurement

Rinse a clean cell (5.3) with the test solution maintained at 20°C, and fill it with this solution. If necessary, use a syringe for filling to avoid any formation of air bubbles.

In the same way, fill the second cell (5.3) with the reference solution, containing only methyl stearate.

Place the cell containing the test portion in one beam of the spectrophotometer (5.2) and the cell containing the reference solution in the other beam.

Plot the transmission (or absorption) curve in the range 1050-900 cm^{-1} (10.0 and 10.8 nm) at optimal instrumental settings.

If isolated *trans*-unsaturated components are present, there will be a transmission minimum (or absorption maximum) at about 970 cm^{-1}. Draw a base-line on the curve; this is the straight line between the absorption minima at around 1000 and 925 cm^{-1} (10.0 and 10.8 nm) (see figure).

8. CALCULATION

For a transmission curve :

$$\text{Absorption } A = \log \frac{PR}{QR}$$

where PR and QR are the lengths measured on the graph.

For an absorption curve :

$$\text{Absorption } A = A_Q - A_P$$

where A_Q is the absorption at Q,

A_P is the absorption at P.

Using the calibration curve constructed according to 7.1.3, determine the amount m', in mg, of methyl elaidate corresponding to the value found for the absorption A.

The per cent amount of isolated *trans*-unsaturated compounds, expressed as methyl elaidate, is:

$$\frac{100m'}{m}$$

where m is the mass, in mg, of the test portion.

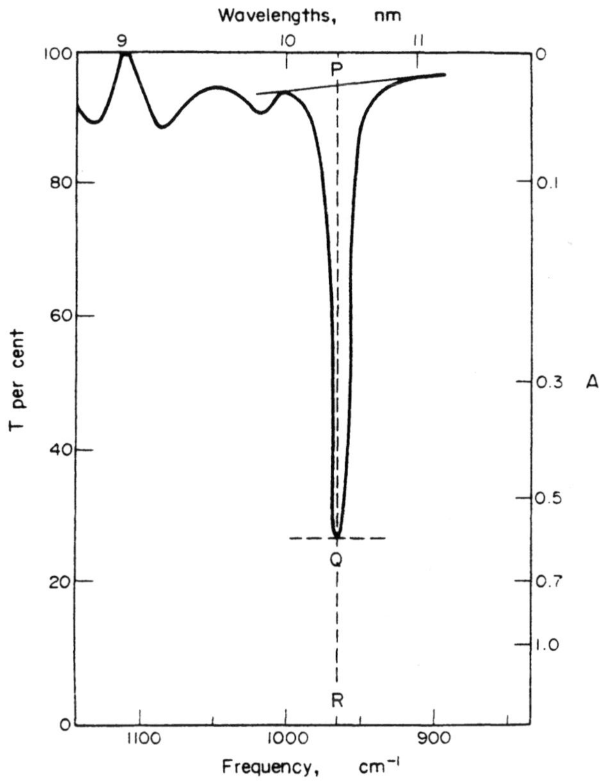
Wavelengths, nm
9
10
11
100
80
60
40
20
0
T per cent
0
0.1
0.3 A
0.5
0.7
1.0
P
Q
R
1100
1000
900
Frequency, cm⁻¹

2.208. Determination of *trans*-Octadecenoic Acids by Thin-Layer Chromatography and Gas-Liquid Chromatography

(Fifth Edition : Method II.D.24)

1. SCOPE

This Standard describes a method for the determination of content of *trans*-octadecenoic acids by thin-layer chromatography and gas-liquid chromatography, and determines the total of elaidic acid and other *trans*-octadecenoic acids.

2. FIELD OF APPLICATION

This Standard is applicable to natural, refined and hydrogenated animal and vegetable oils and fats, fatty acids and fatty acid methyl esters.

It is particularly suitable for the determination of low amounts of *trans* acids.

3. PRINCIPLE

Quantitative conversion of oils, fats and fatty acids into methyl esters, then thin-layer chromatography on silica impregnated with silver nitrate of the methyl esters. Quantitative collection of the band corresponding to the *trans*-octadecenoates and extraction of these in the presence of an internal reference standard. Gas-liquid chromatography of the extracted mixture, and calculation of the content of *trans*-octadecenoates by reference to the amount of the internal standard.

4. APPARATUS

4.1. 250-ml conical flask.

4.2. 1-ml micro beaker.

4.3. Ground-necked test-tubes and extractor as shown in the diagram (*note 1*).

4.4. 1-ml capillary pipette.

4.5. 2-ml graduated pipette.

4.6. Spreader for thin-layer chromatography, inert to silver nitrate.

4.7. Glass plates for thin-layer chromatography, 20 x 20 cm.

4.8. Glass developing tank for thin-layer chromatography with ground-glass lid, suitable for the 20 x 20 cm plates.

4.9. Spray to spray the indicator (5.9) over the plates.

4.10. Desiccator.

4.11. Oven.

4.12. Gas chromatograph, with usual polar column (e.g. diethyleneglycol succinate).

5. REAGENTS

5.1. Light petroleum, b.p. 40 - 60°C, bromine value less than 1.

5.2. Toluene, analytical reagent quality.

5.3. Carbon tetrachloride, analytical reagent quality.

5.4. Diethyl oxide, freshly distilled, free from peroxides and residue.

5.5. Methyl heptadecanoate (or, failing this, methyl myristate).

5.6. Silica powder, with binder, of the usual quality for thin-layer chromatography.

5.7. Silver nitrate, analytical reagent quality, 100 g/l aqueous solution.

5.8. Methyl elaidate, 10 g/l solution in carbon tetrachloride (5.3).

5.9. 2',7'-dichlorofluorescein, 1 g/l solution in 95 per cent (V/V) ethanol.

6. PROCEDURE

6.1. Preparation of the Plates

Thoroughly clean the glass plates (4.7) with ethanol, light petroleum and acetone in order to remove all fatty matter.

Put 30 g of the silica powder with binder (5.6) into a conical flask (4.1). Add 60 ml of the aqueous solution of silver nitrate (5.7). Stopper and shake vigorously. Immediately fill the slurry into the spreader (4.6). Spread as a layer 0.25 mm thick on the cleaned glass plates.

Dry the plates for 15 minutes in the air, then place them in the oven (4.11) at 60°C. Raise the temperature progressively so that it reaches 115°C in 45 minutes. Hold at this temperature for 90 minutes.

Cool and store the plates in a desiccator (4.10) *away from light* and moisture.

6.2. Thin-layer Chromatography

Put about 100 ml of a mixture of toluene (5.2) and light petroleum (5.1), 75/25 (V/V) into the developing tank (4.8), and allow an hour or two for the liquid and vapour to reach equilibrium.

Weigh accurately into the micro beaker (4.2) about 10 mg of the methyl esters under examination, which have been prepared according to 2.301. Add 0.7 to 0.8 ml of carbon tetrachloride (5.3) (*note 2*) to dissolve the test portion.

By means of the capillary pipette (4.4) (*note 3*) drip the *whole* of the above solution in droplets over a length of about 15 cm along the base line of a plate prepared according to 6.1. To one side, near one edge of the plate, put a drop of the reference solution of methyl elaidate (5.8), evaporate the carbon tetrachloride with the help of a stream of nitrogen.

Place the plate in the developing tank (4.8) and, *in the dark,* allow the solvent to migrate until the solvent front is within 1 cm of the top edge of the plate.

Remove the plate from the tank. Allow the solvents to evaporate; this may be accelerated by a current of hot air (e.g. from a hair-dryer). Spray the plate with the dichlorofluorescein solution (5.9). Examine the plate under ultraviolet light. Mark off the band corresponding to the reference standard methyl elaidate, which appears between that of the saturated esters, near the solvent front, and that of methyl oleate (*note 4*).

6.3. Extraction of the Methyl *trans*-Octadecenoates

Scrape off quantitatively the band marked off in 6.2 and transfer it to a ground-necked tube (4.3) as illustrated in the diagram (*note 1*). By means of a graduated pipette (4.5), add a suitable amount (*note 5*) of a solution of the internal standard methyl heptadecanoate (5.5) in diethyl ether (5.4). Make up to 5 ml with diethyl ether. Insert the ground stopper and mix well for 5 minutes by shaking by hand.

Remove the stopper and rinse it with 1 ml of diethyl ether.

Fit the tube at (B) onto the "bridge" (C) of the extractor, and an empty tube at (A). Produce a partial vacuum by connecting the tubulure (E) to a water-pump and draw the fluid from (B) into (A) while filtering it through the filter paper (D).

Repeat the extraction 3 or 4 times, using 5 ml of diethyl ether each time, collecting the extracts in the same receiver at (A). If necessary, partially evaporate under vacuum the solution in tube (A).

6.4. Gas Chromatography

Carry out the gas chromatography of the collected solution from tube (A) in the usual manner (see 2.302). The conditions must be such that the peaks of the *trans*-octadecenoates and of the internal standard are well separated, symmetrical and sufficiently large.

Measure on the chromatogram the areas of the peaks corresponding to the internal standard and to the methyl *trans*-octadecenoates.

7. EXPRESSION OF RESULTS

The content of methyl *trans*-octadecenoates, in per cent (m/m), is given by the formula:

$$\frac{m_1}{m} \times \frac{S}{S_1} \times 100$$

where m is the mass, in mg, of the test portion,

m_1 is the mass, in mg, of the internal standard introduced into the extraction solvent,

S is the area of the peak of the methyl *trans*-octadecenoates,

S_1 is the area of the peak of the internal standard, measured in the same units as S.

8. NOTES

1. The extraction apparatus illustrated in the annexed figure is recommended. It comprises two test tubes, about 125 x 16 mm fitted with 14/23 ground sockets which can be fitted at (A) and (B) onto the "bridge" (C). At one end of the bridge (C), a piece of suitable hardened filter paper (D) is fixed over the end of the tube and clipped in place by a metal band fitted in a groove on the tube. The bridge (C) carries a lateral tubulature enabling it to be connected to a water-pump.

Nevertheless, any other apparatus which will effect a quantitative extraction of the methyl esters may be used, provided it is suitable for the operations described in 6.3.

2. Chloroform may be used instead of carbon tetrachloride (5.3).

3. An automatic dropping apparatus may be used in place of the capillary pipette (4.4).

4. The various methyl *trans*-octadecenoates have the same R_f.

5. According to the expected content of methyl *trans*-octadecenoates, take the following volumes of the diethyl ether (5.4) solutions of known concentration of the internal reference standard (5.5) : this is in order to have the amounts of the methyl *trans*-octadecenoates and the internal standard of the same order of magnitude.

Expected content	Volume to be taken
> 15 per cent	2 ml of a 1 g/l solution
7 to 15 per cent	1 ml of a 1 g/l solution
3 to 7 per cent	0.5 of a 1 g/l solution, or
	5 ml of a 0.1 g/l solution
1 to 3 per cent	2 ml of a 0.1 g/l solution
< 1 per cent	1 ml of a 0.1 g/l solution

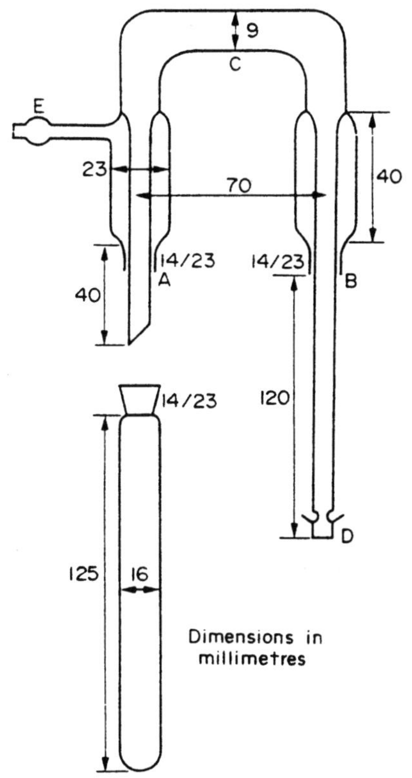

Extraction apparatus

2.210. Determination of Fatty Acids in the 2-Position in the Triglycerides of Oils and Fats

(Fifth Edition : Method II.D.27)

1. SCOPE

This Standard describes a method for the determination of the composition of that fraction of the fatty acids of an oil or fat which is esterified at the 2-position (β, or internal position) of the glycerol.

2. FIELD OF APPLICATION

This Standard is applicable to oils and fats having a melting point below 45°C, owing to the peculiarities of the action of pancreatic lipase.

It is not applicable unreservedly to oils and fats containing substantial amounts of :

fatty acids with twelve or fewer carbon atoms (coconut and palmkernel oils, butterfat), or highly unsaturated fatty acids (with more than four double bounds) containing twenty or more carbon atoms (fish and marine animal oils),

or fatty acids containing oxygenated groups, other than the acid group.

3. PRINCIPLE

Possible neutralization of acid oils and fats in a solvent. Purification by passing onto an alumina column. Partial hydrolysis of triglycerides by pancreatic lipase during a determined time. Separation of the formed monoglycerides by thin-layer chromatography and methanolysis of these monoglycerides. Analysis of these methyl esters by gas-liquid chromatography.

4. APPARATUS

4.1. 100-ml round-bottomed flask.

4.2. 25-ml round-bottomed flask, with ground joint.

4.3. 1-m-long air-condenser, to fit the flask 4.2.

4.4. 250-ml conical flask.

4.5. 50-ml beaker.

4.6. 500-ml separating funnel.

4.7. Chromatographic glass column, 13 mm internal diameter, 400 mm in length, fitted with a fritted glass disc and a tap.

4.8. 10-ml centrifuge tube, with ground glass stopper.

4.9. 5-ml burette, graduated in 0.05 ml.

4.10. 1-ml hypodermic syringe, fitted with a thin needle.

4.11. Microsyringe, to deliver drops of 3-4 µl.

4.12. Spreader for thin-layer chromatography.

4.13. Glass plates for thin-layer chromatography, 20 x 20 cm.

4.14. Glass developing tank for thin-layer chromatography, with ground-glass lid, suitable for the 20 x 20 plates.

4.15. Spray for thin-layer chromatography.

4.16. Oven regulated at 103 ± 2°C.

4.17. Thermostat regulatable between 30°C and 45°C to within 0.5°C.

4.18. Rotary evaporator.

4.19. Vibrating electric shaker, allowing vigorous agitation of the centrifuge tube.

4.20. Ultraviolet lamp for the examination of the thin-layer plates.

And for the control of the lipase activity :

4.21. pH meter.

4.22. Spiral stirrer.

4.23. 5-ml burette.

4.24. Stop-watch.

And for the possible preparation of the lipase :

4.25. Laboratory stirrer, suitable for the dispersion and mixture of heterogeneous materials.

5. REAGENTS

5.1. *n*-Hexane, or, failing this, light petroleum (b.p. 30-50°C), chromatographic quality.

5.2. 2-Propanol, or ethanol, 95 per cent (V/V), analytical reagent quality.

5.3. 2-Propanol, or ethanol, 1/1 aqueous solution.

5.4. Diethyl ether, free from peroxides.

5.5. Acetone.

5.6. Formic acid, at least 98 per cent (m/m).

5.7. Developing solvent : mixture of *n*-hexane (5.1), diethyl ether (5.4) and formic acid (5.6) in proportions 70/30/1 (V/V/V).

5.8. Activated alumina for chromatography, neutral, grade Brockmann I.

5.9. Silica powder, with binder, of suitable quality for thin-layer chromatography.

5.10. Pancreatic lipase of suitable quality (*note 1, note 2*).

5.11. Sodium hydroxide, 120 g/l aqueous solution.

5.12. Hydrochloric acid, aqueous solution 6 N.

5.13. Calcium chloride ($CaCl_2$), 220 g/l aqueous solution.

5.14. Sodium cholate (enzymatic quality), 1 g/l aqueous solution.

5.15. Buffer solution : 1 M aqueous solution of *tris*-hydroxymethylaminomethane bring to pH 8 by addition of hydrochloric acid (5.12) (check by potentiometer).

5.16. Phenolphthalein, 10 g/l solution in 95 per cent (V/V) ethanol.

5.17. 2',7'-Dichlorofluorescein, 2 g/l solution in 95 per cent (V/V) ethanol, rendered slightly alkaline by the addition of 1 drop of 1 N sodium hydroxide solution per 100 ml.

And for the control of the lipase activity :

5.18. Neutralized oil.

5.19. Sodium hydroxide, 0.1 N aqueous solution.

5.20. Sodium cholate (enzymatic quality), 200 g/l aqueous solution.

5.21. Gum arabic, 100 g/l aqueous solution.

6. PREPARATION OF THE SAMPLE

If the sample has an acidity below 3 per cent, determined according to 2.201, purify directly over alumina according to 6.2.

If the sample has an acidity over 3 per cent, determined according to 2.201 : neutralize by alkali in the presence of a solvent according to 6.1, then pass over alumina according to 6.2.

6.1. Neutralization by Alkali in the Presence of Solvent

Into a separating funnel (4.6) introduce about 10 g of the crude oil and add 100 ml of hexane (5.1), 50 ml of 2-propanol (5.2), a few drops of phenolphthalein solution (5.16), and an amount of the sodium hydroxide solution (5.11) corresponding to the free acidity of the oil plus 0.3 per cent excess. Shake vigorously for 1 minute, add 50 ml of distilled water, shake again and leave to settle.

After separation, remove the bottom soap layer. Also remove any intermediate layers (mucilage, insoluble matter). Wash the hexane solution of the neutralized oil with successive 25-30-ml portions of the 2-propanol solution (5.3) until the pink colour of the phenolphthalein disappears.

Remove most of the hexane by distillation under vacuum in the rotary evaporator (4.18), dry the oil at 30-40°C under vacuum with the help of a current of pure nitrogen until the hexane has been removed completely.

6.2. Purification Through Alumina

Prepare a suspension of 15 g of activated alumina (5.8) in 50 ml hexane (5.1) and pour it, while stirring, onto the chromatographic column (4.7). Let the alumina settle evenly, and allow the solvent level to fall to within 1-2 mm above the absorbent. Carefully pour onto the column a solution of 5 g of oil in 25 ml of hexane (5.1); collect the whole of the effluent from the column in a round-bottomed flask (4.1).

7. PREPARATION OF THE CHROMATOGRAPHIC PLATES

Thoroughly clean the glass plates (4.13) with ethanol, light petroleum and acetone to eliminate any trace of fatty matter.

In a conical flask (4.4) place 30 g of silica powder (5.9). Add 60 ml of distilled water. Stopper and shake vigorously for 1 minute. Transfer the slurry immediately to the spreader (4.12) and coat the clean plates with a layer 0.25 mm thick.

Dry the plates in the air for 15 minutes and then for an hour in the oven (4.16) at 103 ± 2°C. Cool the plates in a desiccator to room temperature before use.

Prepared plates are available in commerce.

8. PROCEDURE

8.1. Hydrolysis with Pancreatic Lipase

Into the centrifuge tube (4.8) weigh about 0.1 g of the prepared sample; if the sample is a liquid oil proceed directly as below; if it is a solid fat it must be dissolved in 0.2 ml of hexane (5.1) with gentle warming if necessary.

Add 20 mg of lipase (5.10) and 2 ml of the buffer solution (5.15). Shake well, but carefully, and then add 0.5 ml of the sodium cholate solution (5.14) and 0.2 ml of the calcium chloride solution (5.13). Close the tube with the ground stopper, shake cautiously (avoid wetting the stopper) and put the tube immediately in the thermostat (4.17) maintained at 40 ± 0.5°C and shake by hand for exactly 1 minute.

Remove the tube from the thermostat, and agitate vigorously by means of the electric shaker (4.19) for exactly 2 minutes.

Cool immediately in running water; add 1 ml of hydrochloric acid (5.12) and 1 ml of diethyl ether (5.4). Stopper and mix vigorously by means of the electric shaker. Allow to stand and remove the organic layer by means of the syringe (4.10), if necessary after centrifuging.

8.2. Separation of the Monoglycerides by Thin-Layer Chromatography

Apply the extract to the chromatographic plate with the micro-syringe (4.11), about 1.5 cm from the bottom edge, in a thin, uniform line, as narrow as possible. Set the plate in the well-saturated developing tank (4.14) and develop with the developing solvent (5.7) at about 20°C, up to about 1 cm from the top edge of the plate.

Dry the plate in the air at the temperature of the tank, and spray it with the 2',7'-dichlorofluorescein solution (5.17). Identify the monoglyceride band (R_f about 0.035) under ultraviolet light (4.20).

8.3. Analysis of the Monoglycerides by Gas-Liquid Chromatography

Remove the band obtained in 8.2 with the help of a spatula (avoid removing components remaining on the base line) and transfer in the methylation flask (4.2).

Treat the collected silica directly as described in 2.301, so as to convert the monoglycerides into methyl esters, and then examine the esters by gas chromatography as described in 2.302.

9. EXPRESSION OF RESULTS

Calculate the fatty acid composition in the 2-position to one decimal place (*note 3*).

10. NOTES

1. Checking of the activity of the lipase

Prepare an oil emulsion by shaking a mixture of 165 ml of the solution of gum arabic (5.21), 15 g of crushed ice and 20 ml of a neutralized oil (5.18) in a suitable agitator.

In a beaker (4.5) place 10 ml of this emulsion, followed successively by 0.3 ml of the sodium cholate solution (5.20) and 20 ml of distilled water.

Put the beaker into a thermostat maintained at 37 ± 0.5°C (*note 4*); insert the electrodes of a pH meter (4.21) and a spiral stirrer (4.22).

By means of a burette (4.23) add dropwise the sodium hydroxide solution (5.19) until the pH reaches 8.5.

Add sufficient of an aqueous suspension of the lipase (see below). As soon as the pH meter indicates a pH of 8.3 start the stop-watch (4.24) and drip in the sodium hydroxide solution (5.19) at such a rate as to maintain the pH at 8.3. Read off the volume of alkali solution consumed every minute.

Record the observations in the form of a graph, using the time readings as abscissae and the ml of alkali solution required to maintain the pH constant as ordinates. A linear graph should be obtained.

The lipase suspension mentioned above is a 1 per thousand (m/m) suspension in water. For each test sufficient of this suspension should be used so that about 1 ml of the alkali solution is consumed in 4 to 5 minutes. Usually about 1-5 mg of the powder is required.

The lipase unit is defined as the amount of enzyme which will liberate 10 μ-equivalents of acid per minute. Then the activity A of the powder used, measured in lipase units per mg, is given by the formula :

$$A = \frac{V \times 10}{m}$$

where V is the number of the sodium hydroxide solution (5.19) consumed per minute, calculated from the graph,

m is the mass, in mg, of the test portion of the powder.

2. Preparation of the lipase

Lipases having a satisfactory lipase activity are available commercially. But it is also possible to prepare them in the laboratory as follows :

Chill 5 kg of fresh pig pancreas to 0°C; remove the surrounding solid fat and connective tissue and triturate in a blender so as to obtain a pasty fluid. Stir this paste with the stirrer (4.25) for 4 - 6 hours with 2.5 l of anhydrous acetone and centrifuge. Extract the residue thrice more with the same volume of acetone, then twice with 1/1 (V/V) mixture of acetone and diethyl ether, and twice with diethyl ether.

Dry the residue *in vacuo* for 48 hours to obtain a stable powder, which should be stored in a refrigerator.

3. In every case it is advisable to determine the composition of the total fatty acids of the same sample, since the comparison with that of the acids in the 2-position will help in the interpretation of the figures obtained.

4. The temperature of hydrolysis is set at 37°C, as a liquid oil is being used. However, it is set at 40°C for the test sample, so as to allow the examination of fats with melting points up to 45°C.

2.241. Determination of the Hydroxyl Value (OH.V.)

(Fifth Edition : Method II.D.14)

1. SCOPE

This Standard describes a method for the determination of the hydroxyl value of animal and vegetable oils and fats.

2. FIELD OF APPLICATION

This Standard is applicable to oils and fats containing primary alcohol functions and some secondary alcohols, such as castor oil and monoglycerides. It is not quite exact in the case of fats containing secondary alcohols with a high melting point (hydrogenated castor oils, dihydroxystearic acids).

This Standard is not applicable to fats with an hydroxyl value below 10.

3. DEFINITION

The hydroxyl value is the number of milligrammes of potassium hydroxyde required to neutralize the acetic acid capable of combining by acetylation with 1 g of fat.

4. PRINCIPLE

Acetylation of the fat by a mixture of acetic anhydride and pyridine, in a measured quantity sufficient to provide an excess of reagent.

Conversion of the excess of acetic anhydride into acetic acid and titration with a standardized potasssium hydroxide solution.

5. APPARATUS

5.1. 100-ml round-bottomed flask, or conical flask, with a wide neck.

5.2. Reflux condenser, to fit the flask (5.1).

5.3. 100-ml volumetric flask, with ground-glass stopper.

5.4. 50-ml burette, graduated in 0.1 ml.

5.5. Heating device (water bath, electric hot plate,...).

6. REAGENTS

6.1. Ethanol, 95 per cent (V/V), neutralized.

6.2. Acetic anhydride, analytical reagent quality.

6.3. Pyridine, analytical reagent quality, recently distilled over barium oxide.

6.4. Acetylating reagent : Weigh 25 g of acetic anhydride (6.2) into a volumetric flask (5.3) and make up to the mark with pyridine (6.3). Mix thoroughly by shaking. Store this reagent away from moisture, carbon dioxide and acid vapours in dry, yellow or brown glass-stoppered bottles.

6.5. Potassium hydroxide, 0.5 N solution in ethanol (6.1), accurately standardized.

6.6. Phenolphthalein, 10 g/l solution in ethanol (6.1).

7. PREPARATION OF THE SAMPLE

The hydroxyl value must be determined on a perfectly anhydrous filtered sample.

8. PROCEDURE

Weigh, to the nearest 0.001 g, into the flask (5.1) an amount of the test sample according to Table 1.

Table 1

Expected hydroxyl value	Weight of test portion (g)	Volume of reagent (ml)
10 - 100	2.0	5
100 - 150	1.5	5
150 - 200	1.0	5
200 - 250	0.75	5
250 - 300	0.6	5
	or 1.2	10
300 - 350	1.0	10
350 - 700	0.75	15
700 - 950	0.50	15
950 - 1000	0.30	15
1500 - 2000	0.20	15

Add the appropriate volume of the acetylating reagent (6.4) from a burette (5.4). Fit the reflux condenser (5.2) to the flask.

Place the flask upon the heating device (5.5) and heat it at a temperature of 95-100°C for 1 hour.

Remove the flask from the heating device and allow it to cool.

Add 1 ml of distilled water through the top of the condenser. If the addition of water causes turbidity, add some pyridine (6.3) until clear. Shake. The mixture becomes warm as a result of the conversion of the acetic anhydride into acid.

In order to complete the reaction and decompose any fatty and mixed anhydrides which may have been formed, replace the flask upon the heating device for 10 minutes.

Remove the flask and let it cool to room temperature.

Wash down the condenser and the neck of the flask with 5 ml of ethanol (6.1). Add 2-3 drops of phenolphthalein solution (6.6) and titrate with the standardized potassium hydroxide solution (6.5).

Simultaneously carry out a blank test without the fat under same conditions.

Determine the acid value A.V. of the sample according to 2.201.

9. EXPRESSION OF RESULTS

The hydroxyl value (OH.V.) is given by the formula :

$$\text{OH.V.} = \frac{56.1 \times T \times (V_0 - V)}{m} + \text{A.V.} \quad (\textit{note 1})$$

where V is the number of ml of the standardized ethanolic potassium hydroxide solution (6.5) used for the test portion,

V_0 is the number of ml of the standardized ethanolic potassium hydroxide solution (6.5) used for the blank test,

T is the exact normality of the ethanolic potassium hydroxide solution (6.5) used,

m is the mass, in g, of the test portion,

A.V. is the acid value of the sample.

10. NOTE

1. In the case of highly acid oils the first term in the expression may be negative.

2.251. Determination of Epoxy-Group Oxygen

(Fifth Edition : Method II.D.20)

1. SCOPE

This Standard describes a method for the determination of the amount, as a percentage by mass, of oxygen contained in epoxy-groups present in a fat.

2. FIELD OF APPLICATION

This Standard is applicable to the animal and vegetable oils and fats, except those containing fatty acids with 2,3-epoxy groups, α - β -unsaturated ketones, conjugated dienols, cyclopropenes. It is not applicable to oxidized fats.

3. PRINCIPLE

Direct titration of the fat in benzene solution by hydrogen bromide acetic solution.

4. APPARATUS

4.1. 50-ml conical flasks.

4.2. Burette, the tip of which is fitted through a hole in a rubber stopper which will close the titration flask. The hole in the rubber stopper should allow air to escape during titration.

4.3. Magnetic stirrer with magnetic stirring bars which will not be attacked by the reagents.

5. REAGENTS

5.1. Benzene, or chlorobenzene, analytical reagent quality.

5.2. Glacial acetic acid, free from acetic anhydride.

5.3. Potassium hydrogen phthalate : dry potassium hydrogen phthalate for 2 hours at 120°C, cool and store in a desiccator.

5.4. Hydrogen bromide, about 0.1 N acetic solution, accurately standardized, prepared as follows :

- by bubbling hydrogen bromide gas through acetic acid (5.2),
- or by diluting the commercially available solution of hydrogen bromide in acetic acid.

5.5. Hexamethylpararosaniline hydrochloride (crystal violet), 1 g/l acetic acid (5.2) solution.

6. STANDARDIZATION OF THE SOLUTION OF HYDROGEN BROMIDE IN ACETIC ACID

Weigh accurately, to the nearest 0.1 mg, about 0.4 g of dry potassium hydrogen phthalate (5.3) into a conical flask (4.1). Dissolve it in 10 ml acetic acid (5.2) by carefully warming.

Cool and titrate this solution at room temperature with the hydrogen bromide acetic solution (5.4), using not more than 0.1 ml of the indicator solution (5.5).

The normality T of the reagent is given by the formula :

$$T = \frac{m_o}{0.2042V_o}$$

where V_o is the number of ml of the reagent (5.4),

m_o is the mass, in g, of potassium hydrogen phthalate (5.3) taken for the determination.

This standardization should be carried out in duplicate and the difference between the two determinations must not exceed 0.0004 N.

The normality of the reagent (5.4) must be determinate each day, because it changes on storage.

7. PROCEDURE

Prepare the sample according to 2.001, "Preparation of the sample".

Weigh accurately, to the nearest 0.1 mg, 0.3 to 0.5 g of the prepared sample into a conical flask (4.1). Dissolve it in 10 ml of benzene (5.1). Introduce the magnetic stirring bar (4.3) and add not more than 0.1 ml of the indicator solution (5.5). Fix the stopper and the burette (4.2) in such a manner that the tip of the burette just touches the surface of the solution (in order to avoid loss of hydrogen bromide). Begin stirring and titrate, rapidly at first and then more slowly near the end point, with the hydrogen bromide acetic solution (5.4) until a bluish-green colour persists for 30 seconds (control the speed of the stirrer so as to avoid splashing).

8. EXPRESSION OF RESULTS

The percentage (t) (m/m) of epoxy-group oxygen in the fat is given by the formula :

$$t = \frac{1.60 \times V \times T}{m}$$

where V is the number of ml of the standardized hydrogen bromide acetic solution (5.4) used for the titration,

T is the exact normality of the hydrogen bromide acetic acid solution (5.4) used,

m is the mass, in g, of the test portion.

Section II

Oils and Fats

Determination of the Principal Constituents

2.301. Preparation of the Fatty Acid Methyl Esters

(Fifth Edition : Method II.D.19)

1. SCOPE

This Standard describes some methods for the conversion of animal and vegetable oils and fats, or fatty acids of any source, into the methyl esters of the fatty acids.

The methyl esters so produced can be used in the various (analytical) procedures requiring such derivatives, e.g. gas-liquid chromatography, thin-layer chromatography, infrared spectrophotometry....

2. FIELD OF APPLICATION

The methods described are applicable to animal and vegetable oils and fats and fatty acids from all sources.

The methods given in 3 and 4 are applicable to the preparation of methyl esters of fatty acids containing six or more carbon atoms. If fatty acids containing six and eight carbon atoms are present and for the preparation of methyl esters for gas-liquid chromatography, it is necessary not to evaporate the solvent of the solution of methyl esters.

The general method 3 may lead to erroneous results in the following cases :

- compounds having secondary oxygen groupings (hydroxy, hydroperoxy, keto, epoxy),
- compounds containing cyclopropan- and cyclopropene-groups,
- conjugated polyunsaturated compounds and acetylenic compounds,
- waxes.

For these it is preferable to use one of the methods described under 4. Nevertheless, if the fatty matter contains such compounds only in very small amounts (e.g. cottonseed oil) it may be esterified according to 3.

Phospholipids may be analysed after saponification and esterification of the fatty acids.

Unsaponifiable matter is not removed and, if present in a substantial amount, it may interfere with subsequent analyses (*note 1*).

3. GENERAL METHOD FOR THE PREPARATION OF METHYL ESTERS OF FATTY ACIDS CONTAINING SIX OR MORE CARBON ATOMS

The general method using boron trifluoride should be used preferentially. If it is impossible, alternative methods are given in 4.

3.1. Principle

Saponification of the glycerides, and liberation and esterification by methanol of the fatty acids in the presence of boron trifluoride.

3.2. Apparatus

3.2.1. 50- and 100-ml ground-necked round-bottomed flasks.

3.2.2. Reflux condenser, 20 to 30 cm effective length, with ground joint appropriate to the flask (3.2.1).

3.2.3. 250-ml separating funnels.

3.2.4. Inlet tube for passing nitrogen.

3.2.5. Test tubes with ground glass stoppers.

3.2.6. Graduated pipette, capacity at least 10 ml, fitted with a rubber bulb, or automatic pipette.

3.2.7. Boiling chips (fat-free).

3.3. Reagents

3.3.1. Heptane, chromatographic quality (*note 2, note 4*).

3.3.2. Redistilled light petroleum (b.p. 40-60°C), bromine value less than 1, residue free, or hexane (*note 2*).

3.3.3. Sodium sulphate, anhydrous.

3.3.4. Sodium hydroxide, approximately 0·5 N methanolic solution :

dissolve 2 g of sodium hydroxide in 100 ml methanol containing not more than 0.5 per cent (m/m) of water. When the solution has to be stored for a considerable time, a small amount of white precipitate of sodium carbonate may be formed; this has no effect on the preparation of the methyl esters.

3.3.5. Boron trifluoride, 12 to 25 per cent (m/m) methanolic solution. 14 and 50 per cent solutions are available commercially (*note 2*).

Caution : Boron trifluoride is *poisonous*. For this reason, it is not recommended that the analyst prepare the methanolic solution of boron trifluoride from methanol and boron trifluoride (*note 3*).

3.3.6. Sodium chloride, saturated aqueous solution.

3.3.7. Methyl red, 1 g/l solution in 60 per cent (V/V) ethanol.

3.3.8. Nitrogen, containing less than 5 mg/kg of oxygen.

3.4. Procedure

Because of the toxic character of boron trifluoride, the following operations are best performed under a ventilated hood. All glassware must be washed with water *immediately after use*.

Prepare the sample according to 2.001, "Preparation of the sample".

Precise weighing is not necessary. The size of the sample is only required in order to select the appropriate size of flask and the quantities of the reagents, according to the following table :

Weight of sample (mg)	Flask capacity (3.2.1) (ml)	0.5 NaOH (3.3.4) (ml)	Methanolic solution of BF_3 (3.3.5) (ml)
100 - 250	50	4	5
250 - 500	50	6	7
500 - 750	100	8	9
750 - 1000	100	10	12

If the methyl esters are intended for an analysis by gas-liquid chromatography, a sample of about 350 mg is preferred. If it is smaller, care should be taken to ensure that sample is a representative one (*note 5*).

If the oil or fatty acids include fatty acids containing more than two double bonds, it is advisable to purge the air from the methanol and the flask by passing a stream of nitrogen into the methanol for a few minutes.

3.4.1. For oils and fats (note 1)

Introduce the prepared sample into the prescribed flask (3.2.1). Add the appropriate amount of methanolic sodium hydroxide solution (3.3.4) and a boiling chip (3.2.7). Fit the condenser (3.2.2) to the flask. Boil under reflux until the droplets of fat disappear (this may take 5 to 10 minutes, but in certain exceptional cases it may require more than 10 minutes).

Add the appropriate amount of methanolic boron trifluoride solution (3.3.5) from the bulb or automatic pipette (3.2.5) through the top of the condenser, and continue boiling for 1 minute.

Add 2 to 5 ml of heptane (3.3.1) (*note 4*) (the precise amount does not affect the reaction) to the boiling mixture through the top of the condenser and continue boiling for 1 minute.

Withdraw the source of heat, and then remove the condenser. Add a small portion of saturated sodium chloride solution (3.3.6) and shake the flask gently by rotating it several times.

Add more saturated sodium chloride solution to the flask in order to bring the level of liquid into the neck of the flask. Allow to separate and transfer about 1 ml of the upper layer (heptane solution) into a ground-necked test tube (3.2.5) and add a little anhydrous sodium sulphate (3.3.3) to remove any trace of water. If the sample taken was 350 mg, this solution contains about 7-17 per cent of methyl esters and may be injected directly onto the column for gas-liquid chromatography. In the other cases dilute the heptane solution to obtain a 5-10 per cent concentration of methyl esters (*note 6*).

In order to recover the whole of the dry esters, transfer the saline solution and the heptane layer into a separating funnel (3.2.3). Separate the layers. Extract the saline solution twice with 50-ml portions of light petroleum (3.3.2). Combine the heptane solution and the two extracts, and wash them with 20-ml portions of water until free from acid (methyl red indicator (3.3.7.)). Dry over anhydrous sodium sulphate, filter and evaporate the solvent over a boiling-water bath in a current of nitrogen (3.3.8) (*note 6, note 7*). For samples under 500 mg it is desirable to reduce proportionately the volumes of solvent and water used.

3.4.2. For fatty acids

If the sample consists entirely of fatty acids, the saponification step is not necessary.

Place the desired amount of prepared fatty acids in the appropriate flask (3.2.1). Add the appropriate amount of methanolic boron trifluoride solution (3.3.5). Boil for 2 minutes. Then proceed as in 3.4.1 from "Add to the boiling mixture...".

4. ALTERNATIVE METHODS NOT INVOLVING THE USE OF BORON TRIFLUORIDE

4.1. For Neutral Oils and Fats (Acid Value <2)

4.1.1. Principle

Methanolysis of the glycerides in an alkaline medium.

4.1.2. Apparatus

4.1.2.1. 100-ml ground-necked conical or round-bottomed flasks.

4.1.2.2. Reflux condenser with ground joint appropriate to the flask (4.1.2.1).

4.1.2.3. 50-ml narrow-mouthed conical flask.

4.1.2.4. 125-ml or, failing this, 250-ml separating funnel.

4.1.2.5. Inlet tube for passing nitrogen.

4.1.2.6. High-speed stirrer, with appropriate heating means (e.g. magnetic stirrer with a heater).

4.1.2.7. Boiling chips (fat-free).

4.1.3. Reagents

4.1.3.1. Heptane, chromatographic quality (*note 2, note 4*).

4.1.3.2. Methanol, containing not more than 0.5 per cent (m/m) of water.

4.1.3.3. Sodium sulphate, anhydrous.

4.1.3.4. Potassium hydroxide, approximately 1 N methanolic solution : dissolve 5·6 g of potassium hydroxide in 100 ml of methanol (4.1.3.2).

4.1.3.5. Nitrogen, containing less than 5 mg/kg of oxygen.

4.1.4. Procedure (note 1)

Prepare the sample according to 2.001, "Preparation of the sample".

Transfer about 4 g (*note 5*) of the prepared fat into a round-bottomed or conical flask (4.1.2.1).

If the oils include fatty acids containing more than two double bonds, it is advisable to purge the air from the methanol and the flask by passing a stream of nitrogen into the methanol for a few minutes.

Add about 40 ml of methanol (4.1.3.1), 0.5 ml of the potassium hydroxide solution (4.1.3.4) and a boiling chip (4.1.2.7). Fit the reflux condenser (4.1.2.2), stir and bring to boil. The solution should become clear. The reaction is usually complete after 5 to 10 minutes (*note 8*).

Withdraw the source of heat, and then remove the condenser. Cool the flask under running water and transfer the contents to a separating funnel (4.1.2.4). Rinse the flask with 20 ml of heptane (4.1.3.1) (*note 4*) and transfer it into the separating funnel. Add about 40 ml of water, shake and allow to separate. The esters pass into the upper heptane layer. Separate. Extract the aqueous layer again with 20 ml of heptane. Combine the two extracts and wash them with 20-ml portions of water. Separate and dry the ester solution over anhydrous sodium sulphate (4.1.3.3). Filter through cotton wool into a conical flask (4.1.2.3) and evaporate the solution down to approximately 20 ml over a boiling-water bath, while passing a stream of nitrogen (4.1.3.5) (*note 6, note 7*).

4.2. For Acid Oils and Fats (Acid Value >2) and Fatty Acids

4.2.1. Principle

For acid oils and fats, neutralization of the free fatty acids and alkaline methanolysis of the glycerides followed by esterification of the fatty acids in acid medium.

4.2.2. Apparatus

4.2.2.1. 250-ml ground-necked conical or round-bottomed flask.

4.2.2.2. Reflux condenser with ground joint appropriate to the flask (4.2.2.1).

4.2.2.3. 250-ml separating funnels.

4.2.2.4. 100-ml narrow-necked conical flask.

4.2.2.5. Inlet tube for passing nitrogen.

4.2.2.6. Boiling chips (fat-free).

4.2.3. Reagents

4.2.3.1. Heptane, chromatographic quality (*note 2, note 4*).

4.2.3.2. Sodium sulphate, anhydrous.

4.2.3.3. Sodium methylate, methanolic solution:
Dissolve 1 g of sodium in 100 ml of methanol containing not more than 0.5 per cent (m/m) of water.

4.2.3.4. Hydrochloric acid, approximately 1 N anhydrous methanolic solution (*note 9, note 10*).

4.2.3.5. Methyl red, 1 g/l solution in 60 per cent (V/V) ethanol.

4.2.3.6. Nitrogen, containing less than 5 mg/kg of oxygen.

4.2.4. Procedure for use with acid oils and fats (note 1)

Prepare the sample according to 2.001, "Preparation of the sample".

Transfer about 4 g (*note 5*) of the prepared fat into a conical or round-bottomed flask (4.2.2.1).

If the acid oils include fatty acids containing more than 2 double bonds, it is advisable to purge the methanol and the flask by passing a stream of nitrogen into the methanol for a few minutes.

Add 40 ml of the sodium methylate solution (4.2.2.3) (*note 11*) and a boiling chip (4.2.2.6). Fit the reflux condenser (4.2.2.2), stir and bring to the boil. The solution should become clear, which usually occurs in about 10 minutes. The reaction is normally complete after 15 minutes (*note 8*).

Then add at least 50 ml of the methanolic hydrochloric acid solution (4.2.3.4) to the flask and boil again for 10 minutes (*note 12*).

Withdraw the source of heat, and then remove the condenser. Cool the flask under running water, add 100 ml of water to the flask, then pour the contents into a separating funnel (4.2.2.3) and 30 ml of heptane (4.2.3.1) (*note 4*). Shake vigorously and leave to settle until the two phases have separated. Collect the heptane layer. Extract the aqueous phase again with 30 ml of heptane. Combine the two heptane extracts and wash them several times with water until neutral (methyl red indicator (4.2.3.5)) and dry over anhydrous sodium sulphate (4.2.3.2). Filter through cotton wool into a conical flask (4.2.2.4) and finally evaporate the solution down to 20 ml over a boiling-water bath, while passing a stream of nitrogen (4.2.3.6) (*note 6, note 7*).

4.2.5. Procedure for use with fatty acids (note 1)

The following simplified procedure is suitable for fatty acids.

Prepare the sample according to 2.001, "Preparation of the sample".

Transfer about 4 g (*note 5*) of the prepared fatty acids into a conical or round-bottomed flask (4.2.2.1).

If the fatty acids include fatty acids containing more than two double bonds, it is advisable to purge the air from the methanol and the flask by passing a stream of nitrogen (4.2.3.6) into the methanol for a few minutes.

Add 50 ml of the hydrochloric acid solution (4.2.3.4) and a boiling chip (4.2.2.6). Fit the condenser and boil for 10 minutes under reflux.

Withdraw the source of heat, and then remove the condenser. Then operate as in 4.2.4 from "Cool the flask under running water...".

5. SPECIAL METHOD FOR THE PREPARATION OF THE METHYL ESTERS OF FATTY ACIDS CONTAINING FOUR OR MORE CARBON ATOMS

5.1. Field of Application

The method is principally destine to the preparation of methyl esters for analysis by gas-liquid chromatography.

The method is not suitable for acid oils and fats (acid value >2).

5.2. Principle

Methyl esters are formed by the reaction of a fat with methanolic potassium hydroxide as an intermediate stage before the saponification takes place.

For the quantitative determination by gas-liquid chromatography of the lower fatty acids with four and six carbon atoms, an internal standard (methyl pentanoate) must be added to the methyl esters.

For the determination by gas-liquid chromatography of the fatty acid composition of the fat and for the other physical determinations, an internal standard is not required.

5.3. Apparatus

5.3.1. Approx. 20-ml test tubes with glass stoppers.

5.3.2. 50- and 100-ml volumetric flasks.

5.3.3. 1-ml (or more) graduated pipettes.

5.3.4. 10-ml measuring cylinders.

5.4. Reagents

5.4.1. Heptane, chromatographic quality (*note 2, note 4*).

5.4.2. Potassium hydroxide, approximately 2 N methanolic solution : dissolve 11.2 g of potassium hydroxide in 100 ml of methanol containing not more than 0.5 per cent (m/m) of water.

5.4.3. Reference solution I : in a 50-ml volumetric flask (5.3.2) weigh to the nearest 0.1 mg about 1 g of methyl pentanoate and make up to the mark with heptane (5.4.1).

5.4.4. Reference solution II : in a 100-ml volumetric flask (5.3.2) weigh to the nearest 0.1 mg about 200 mg of methyl pentanoate and make up to the mark with heptane (5.4.1).

5.5. Procedure

If the quantitative determination of butyric acid by gas-liquid chromatography is required, it is necessary to use a reference solution : reference solution I (5.4.3) if the presumed amount of butyric acid is higher than 1 per cent, reference solution II (5.4.4) if this is lower than 1 per cent.

Prepare the sample according to 2.001, "Preparation of the sample".

In a test tube (5.3.1) weigh to the nearest 1 mg 1 g of the prepared sample. Add 10 ml of heptane (5.4.1) from a measuring cylinder (5.3.4). Add 1.0 ml of the appropriate reference solution from the graduated pipette (5.3.3).

Note : If the determination of butyric acid is not required, it is not necessary to weigh the sample to an accuracy of 1 mg nor to add the reference solution.

Add 0.5 ml of the methanolic potassium hydroxide solution (5.4.2) and mix the contents of the test tube by inversions until the solution becomes clear. This requires about 20 seconds. Almost immediately after clearing, the solution again becomes turbid owing to the separation of glycerol. Settling of the glycerol follows quickly.

Decant the upper layer containing the methyl esters.

The methyl esters may be stored in a refrigerator for several days.

6. NOTES

1. If the unsaponifiable matter interferes, dilute the saponified solution with water and eliminate the unsaponifiable matter by extraction with diethyl ether or hexane as described in 2.401. Acidify the aqueous soap solution and separate the fatty acids. Prepare the methyl esters from these as described in 3.4.2 or 4.2.5.

2. In the course of the gas-liquid chromatography of the methyl esters, certain reagents, particularly the methanolic boron trifluoride solution may produce adventitious peaks on the graph (in the region of C_{20} – C_{22} esters in the case of methanolic boron trifluoride solutions). Consequently any new batch of reagent should be checked by preparing the methyl esters of pure oleic acid, and chromatographing them; if an extraneous peak appears, the reagent should be rejected. The various reagents must not give peaks interfering with those of methyl esters of fatty acids during the gas-liquid chromatography.

   The methanolic solutions of boron trifluoride must be stored in a refrigerator.

3. If it is absolutely unavoidable to prepare a solution of boron trifluoride from gazeous boron trifluoride, the recommended method is :

   Weigh a 2-l flask containing 1 l of methanol. Cool in an ice bath and with the flask still in bath, bubble BF_3 from a cylinder through a glass tube into the methanol until 125 g BF_3 is absorbed. Perform the operation in a fume-cupboard. BF_3 must be flowing through the glass tube before it is placed in and until it is removed from methanol to prevent liquid from being drawn into the gas cylinder valve system. Gas should not flow so fast that white fumes emerge from flask.

   Reagent is stable for 2 years.

4. If fatty acids containing twenty or more carbon atoms are absent, hexane may be substituted for heptane (mixture of pure C_7 isomers tested by gas-liquid chromatography).

5. If the suggested amount of sample is not available, 10 mg, or even less, may be used, provided that the amounts of reagents and the size of the containers are reduced proportionally.

6. Preferably, the solutions of methyl esters should be analysed as soon as possible. If necessary, the heptane solution containing the methyl esters may be stored under an inert gas in a refrigerator. In the case of prolonged storage, it is desirable to protect the methyl esters from autoxidation by adding to the solution an antioxidant in such a concentration as will not interfere with the subsequent analysis, e.g. 0.05 g/l of BHT (2,6-di-*tert*butyl 4-methyl phenol).

   If necessary, the dry and solvent-free methyl esters may be stored 24 hours under inert gas in a refrigerator, or longer in sealed tube under vacuum in a deep-freeze.

7. There is some risk of losing part of the most volatile methyl esters if the evaporation of the solvent is prolonged, or if the current of nitrogen is too vigorous.

   For infrared spectroscopy, elimination of the solvent must be as complete as possible. For gas-liquid chromatography, do not remove solvent.

8. With oils such as castor oil, it is possible that the limpidity is not the test of reaction.

9. Small amounts of gazeous hydrochloric acid can easily be prepared in the laboratory by simple displacement from the commercial solution (ρ = 1.18) by dripping conc. sulphuric acid (ρ = 1.84). The liberated gas is easily dried by bubbling through conc. sulphuric acid. Since hydrochloric acid is very rapidly absorbed by methanol, it is advisable to take the usual precautions in dissolving it, e.g. introduce the gas through a small inverted funnel with the rim just touching the surface of the liquid. Large quantities of methanolic hydrochloric acid solution can be prepared in advance, as it keeps perfectly in glass-stoppered bottles stored in the dark.

10. Approx. 1 N methanolic sulphuric acid may be used in place of the methanolic hydrochloric acid, but esterification requires at least 20 minutes, and the precipitated sodium sulphate hinders boiling, so that either filtration or the use of a magnetic stirrer is necessary.

11. Before introducing the sample, it is quite feasible to place 40 ml of methanol in the flask and to add 0.4 g of sodium, thus preparing an *ad hoc* solution of sodium methylate.

12. In the case of very acid oils and fats, owing the relatively high quantity of sodium methylate, precipitation of sodium chloride occurs, which may lead to bumbling during the subsequent boiling. This precipitate may be filtered off, although, owing to the short period of boiling prescribed, this is usually unnecessary.

2.302. Gas-Liquid Chromatography of Fatty Acid Methyl Esters

(Fifth Edition : Method II.D.25)

1. SCOPE

This Standard is intended to give general guidance for the application of gas-liquid chromatography to determine the qualitative and quantitative composition of a mixture of fatty acid methyl esters.

2. FIELD OF APPLICATION

This Standard is applicable to the fatty acid methyl esters obtained according to 2.301.

The conditions specified in the method are not suitable for determining fatty acids that have been polymerized.

3. APPARATUS

The instructions given relate to the ordinary equipment used for gas-liquid chromatography, employing a packed column and a flame-ionization detector (*note 1*). Any apparatus giving the efficiency and the resolution described in 5.1.2 is suitable.

3.1. Gas-Liquid Chromatography

3.1.1. Injection system

The injection system should have the least dead space possible. If possible, it should be heated to a temperature 20 to 50°C higher than that of the column.

3.1.2. Oven

The oven should be capable of heating the column to at least 220°C and of maintaining the desired temperature to within 1°C.

If programmed heating is to be employed, an apparatus with a twin column is recommended.

3.1.3. Packed column

3.1.3.1. Column. The column must be constructed of a material inert to the substances to be analysed: glass or, failing this, stainless steel (*note 2*).

Length : 1 to 3 m. A relative short column should be used when long-chain acids (>C_{20}) are present. For the determination of C_4 and C_6 fatty acids, a 2-m column is recommended.

Internal diameter : 2 to 4 mm.

3.1.3.2. Packing

Support : acid-washed and silanized diatomaceous earth, or other suitable inert support with a narrow range (25 μm) of grain size between the limits 125-200 μm, the average grain size being related to the internal diameter and length of the column.

Stationary phase : polyester type of polar liquid (e.g. diethylene glycol polysuccinate, butanediol polysuccinate, ethylene glycol polyadipate...) or any other liquid (e.g. cyanosilicones...) meeting the requirements below. The stationary phase should amount to 5 to 20 per cent of the packing. A non-polar stationary phase can be used for certain separations.

3.1.3.3. Conditioning the newly prepared column. The column being disconnected from the detector, if possible, heat the oven gradually to 185°C and pass a current of inert gas through the freshly prepared column at a rate of 20 - 60 ml/min for at least 16 hours at this temperature, and for 2 hours more at 195°C.

3.1.4. Detector

For preference, the detector should be capable of being heated to a temperature above that of the column.

The manipulations described below relate to the use of a flame-ionization detector (*note 1*).

3.2. Syringe

Syringe, maximum capacity 10 μl, graduated in 0.1 μl.

3.3. Recorder

If the recorder curve is to be used to calculate the composition of the mixture analysed, an electronic recorder of high precision is required. It should be compatible with the apparatus used. The characteristics of the recorder should be :

- Rate of response below 1.5 seconds, preferentially below 1 second (the rate of response is the time taken for the recording pen to pass from 0 to 90 per cent following the momentary introduction of a 100 per cent signal).
- Breadth of the paper : 25 cm minimum.
- Paper speed : 25-100 cm/hr.

3.4. Integrator or Calculator (Optional)

Rapid and accurate calculation can be performed with the help of an electronic integrator or calculator. This must give a linear response with adequate sensitivity, and the correction for deviation of the base-line must be satisfactory.

4. REAGENTS

- Carrier gas : inert gas (nitrogen, helium, argon...) thoroughly dried and containing less than 10 mg/kg of oxygen.
- Auxiliary gases : hydrogen (99.9 per cent min.) free from organic impurities, air, or oxygen, free from organic impurities.
- Reference standards : a mixture of methyl esters, or the methyl esters of an oil, of known composition, preferably similar to that of the fatty matter to be analysed.

5. PROCEDURE

5.1. Conditions of Test

5.1.1. Determining optimal operating conditions

In selecting the test conditions, the following variables must be taken in account:

- length and diameter of the column,
- nature and amount of the stationary phase,
- temperature of the column,
- carrier gas flow,
- resolution required,
- size of the sample for analysis,
- time of analysis.

The size of the sample should be chosen so that the assembly of detector and electrometer gives a linear response.

As a rule, the figures shown in Tables 1 and 2 will lead to the desired results, viz. at least 2000 theoretical plates for methyl stearate and its elution within about 15 minutes.

Table 1

Internal diameter of column	Carrier gas supply
2 mm	15-25 ml/min
3 mm	20-40 ml/min
4 mm	40-60 ml/min

Table 2

Concentration of stationary phase	Temperature
5 per cent	175°C
10 per cent	180°C
15 per cent	185°C
20 per cent	185°C

When the apparatus allows it, the injector should be at a temperature of about 200°C and the detector at a temperature equal to, or higher than, that of the column.

The flow of hydrogen to the flame-ionization detector is, as a rule, about half that of the carrier gas, and the flow of oxygen about 5 to 10 times that of the hydrogen.

5.1.2. Determining the efficiency and the resolution

Carry out the analysis of a mixture of methyl stearate and methyl oleate in about equivalent proportions (e.g. methyl esters from cocoa butter). Choose the size of the sample, the temperature of the column and the carrier gas flow so that the maximum of the methyl stearate peak is recorded about 15 minutes after the solvent peak, and rises to three-quarters of the full scale.

Calculate the number of theoretical plates n (efficiency) by the formula :

$$n = 16 \left(\frac{dR_1}{\omega_1} \right)^2$$

and the resolution R by the formula :

$$R = \frac{2\,\Delta}{\omega_1 + \omega_2}$$

where dR_1 is the retention distance, measured in mm, from the start to the maximum of the peak for methyl stearate,

ω_1 and ω_2 are the widths, in mm, of the peaks for methyl stearate and methyl oleate, measured between the points of intersection of the tangents at the inflexion points of the curve with the base line,

Δ is the distance, in mm, between the respective peak maxima for methyl stearate and oleate.

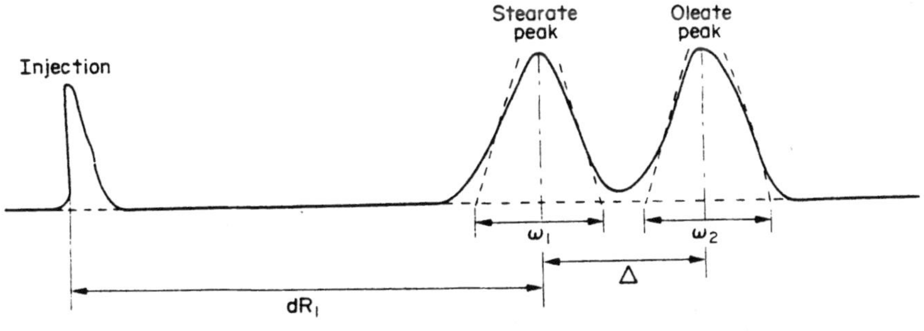

Operating conditions to be selected are those which will afford at least 2000 theoretical plates for methyl stearate, and a resolution of at least 1.25. Additionally, linolenic acid ($C_{18:3}$) ester should be separable from arachidic acid ($C_{20:0}$) and gadoleic acid ($C_{20:1}$) esters.

5.2. Analysis

The sample for examination shall be 0.1 to 2 µl of the heptane solution of methyl esters obtained according to 2.301. In the case of esters not in solution, prepare an approx. 10 per cent solution in heptane and inject 0.1 to 1 µl of this.

If the search is for constituents present only in trace amounts, the sample size may be increased (up to tenfold).

As a rule, the operating conditions will be those defined in 5.1.1. Nevertheless, it is possible to work with a lower column temperature where the determination of acids below C_{12} is required, or at higher temperature when determining fatty acids above C_{20}.

On occasion, it is possible to employ temperature programming in both the previous cases. If the sample contains the methyl esters of fatty acids below C_{12}, it is necessary to inject the sample at 100°C (or at 50-60°C if butyric acid is present) and immediately to raise the temperature at a rate of 4-8°C/min to the optimum. In some cases the two procedures can be combined : after the programmed heating, continue the elution at a constant temperature until all the components have been eluted. If the instrument does not operate with programmed heating, work at two fixed temperatures between 100°C and 195°C.

If possible, it is recommended that analysis be made on two columns of different polarity to check for the absence of coincident peaks, for example in the case of fish oils, or in the case of the simultaneous presence of $C_{18:3}$ and $C_{20:0}$ or of $C_{18:3}$ and conjugated $C_{18:2}$.

6. EXPRESSION OF RESULTS

6.1. Qualitative Analysis

Analyse the reference standard mixture of known composition in the same operating conditions as those employed for the sample, and measure the retention distances (or retention times) for the constituent fatty esters. Using a semi-logarithmic paper, construct the graphs showing the logarithm of the retention distance (or retention time) as a function of the number of carbon atoms of the acids; in isothermal conditions the graphs for straight chain esters of the same degree of unsaturation should be straight lines. These straight lines are approximately parallels.

Identify the peaks for the sample from these graphs, if necessary by interpolation.

It is necessary to avoid conditions such that "masked peaks" exist, i.e. where the resolution is insufficient to separate two components.

6.2. Quantitative Analysis

6.2.1. Determination of the composition

Apart from exceptional cases, use the method of internal normalization, i.e. assume that the whole of the components of the sample are represented on the chromatogram, so that the total of the areas under the peaks represents 100 per cent of the constituents (total elution).

If the equipment includes an integrator, use the figures obtained therefrom. If not, determine the area under each peak by triangulation (multiply the height by the breadth at mid-height) and, where necessary, take into account the various attenuations used during the recording.

6.2.1.1. The general case. If significant amounts of components below C_8 are absent, calculate the content of a given constituent (expressed as a percentage by mass of methyl esters) by determining the percentage represented by the area of the corresponding peak relative to the sum of the areas of all the peaks :

$$\text{per cent (m/m) of the component } i \text{ expressed as methyl ester} = \frac{A_i}{\Sigma A_i} \times 100$$

where A_i is the area of the peak corresponding to component i,

ΣA_i is the sum of the areas under all the peaks.

6.2.1.2. Use of correction factors. In certain cases, particularly in the presence of acids below C_8 or of acids with secondary groups, and with a catharometer detector, correction factors must be used to convert the percentages of peak areas into mass-percentages of the components.

Determine the correction factors with the help of a chromatogram derived from the analysis of a reference mixture of methyl esters of known composition under operating conditions identical with those used for the sample. For this reference mixture :

$$\text{per cent (m/m) of component } i = \frac{m_i}{\Sigma m_i} \times 100$$

where m_i is the mass, in mg, of component i in the reference mixture,

Σm_i is the total of the masses, in mg, of the various components of the reference mixture.

From the chromatogram of the reference mixture, calculate :

$$\text{per cent (area/area) of component } i = \frac{B_i}{\Sigma B_i} \times 100$$

where B_i is the area under the peak corresponding to component i,

ΣB_i is the sum of the areas under all the peaks.

Whence :

$$\text{correction factor } K_i = \frac{m_i \times \Sigma B_i}{B_i \times \Sigma m_i}.$$

Commonly, the correction factors are given in relative factors reported to the correction factor of palmitic acid: K_{16} :

$$K'_i = \frac{K_i}{K_{16}}.$$

Then in the sample, the content of each component is given by :

per cent (m/m) of component i, expressed as methyl ester =

$$\frac{K'_i \times A_i}{\Sigma(K'_i \times A_i)} \times 100.$$

6.2.1.3. *Use of an internal standard.* In certain cases (notably the assay of C_4 and C_6 acids, and the determination of the acids when all the fatty acids are not eluted) an internal standard should be used (C_5 and C_{15} or C_{17} respectively) and the correction factor for the internal standard should be determined.

per cent (m/m) of component i expressed as methyl ester =

$$\frac{m_S \times K'_i \times A_i}{m \times K'_S \times A_S} \times 100$$

where m_S is the mass, in mg, of the internal standard,

m is the mass, in mg, of the test portion,

K'_S is the correction factor for the internal standard,

K'_i is the correction factor for the component i,

A_S is the area of the peak corresponding to the internal standard,

A_i is the area of the peak corresponding to component i.

6.2.2. *Expression of results*

Give the results to 3 significant figures for contents over 10 per cent,
to 2 significant figures for contents between 1 and 10 per cent,
to 1 significant figure for contents below 1 per cent,

i.e. with one figure beyond the decimal point in every case.

6.2.3. *Repeatability*

The difference between the results of two determinations carried out on the same day by the same operator using the same apparatus for the same test material and for constituents present in excess of 5 per cent should not exceed a relative figure of 3 per cent of the determined value, with an absolute maximum of 1 per cent. For components present in amounts of less than

5 per cent, the repeatability in relative terms diminishes progressively as the content is reduced.

6.2.4. Reproducibility

The difference between the results obtained in two different laboratories on the same test material for constituents present in excess of 5 per cent should not exceed a relative figure of 10 per cent of the determined value, with an absolute maximum of 3 per cent. For constituents present in amounts less than 5 per cent, the reproducibility in relative terms diminishes progressively as the content is reduced.

Note : With "test material" the oil, fat or fatty acid sample is indicated and repeatability and reproducibility cover the preparation of the methyl esters according to 2.301 together with the gas-liquid chromatographic analysis described above.

7. NOTES

1. A gas-liquid chromatograph employing a catharometer (working on thermal conductivity changes) may be used. Operating conditions must then be modified as follows :

Column : length : 2 to 4 m,
internal diameter : 4 mm.

Support : grain size between 160 and 200 μm.
Stationary phase : 15 to 25 per cent.
Carrier gas : helium, or, failing this, hydrogen, with a content of oxygen as low as possible.
No auxiliary gases.
Injector temperature : 40 to 60°C above that of the column.
Column temperature : 180 to 200°C.
Flow of carrier gas : usually between 60 and 80 ml/min.
Amount of sample injected : usually between 0.5 and 2 μl.
Quantitative analysis : the correction factors described in 6.2.1.2 must be used.

2. If polyunsaturated components with more than three double bonds are present, they may decompose in a stainless-steel column.

2.321. Determination of Mono-, Di- and Triglycerides by Column Chromatography

(Fifth Edition : Method II.C.7)

1. SCOPE

This Standard describes a method for the determination of mono-, di- and triglycerides in oils and fats by column chromatography.

2. FIELD OF APPLICATION

This Standard is applicable to animal and vegetable oils and fats containing less than 2 per cent of free fatty acids. Samples with 2 to 5 per cent of free fatty acids may be analysed according to the procedure given; in this case, however, the free fatty acid content should be determined in all fractions obtained. More than 5 per cent of free fatty acids may lead to erroneous results by interfering with the chromatographic fractionation.

3. PRINCIPLE

Absorption of the test portion onto silica gel. Elution of tri-, di- and monoglycerides in that order with benzene, benzene containing 10 per cent diethyl oxide and diethyl oxide respectively, and recovered by evaporation. The method given, as all methods for column chromatography, must be regarded as indicative only.

It is, therefore, necessary to control the efficacity and the purity of each batch of silica gel, the purity of the solvents used and to check the fractions obtained by thin-layer chromatography.

4. APPARATUS

4.1. 50- and 500-ml round-bottomed flasks.

4.2. 100- and 150-ml beakers.

4.3. 150-ml conical glass-stoppered flask.

4.4. Chromatographic column, 19 mm diameter and about 300 mm long, equipped with a 19/26 standard socket at the top and a coarse fritted glass disc and a 19/26 standard cone at the bottom (if 19/26 standard ground joints are unavailable, a suitable alternative may be used).

4.5. 250-ml separating funnel, provided with a "Teflon" stopcock and a 19/26 standard cone with angle-drip tip.

4.6. Adapter, consisting of a 19/26 standard socket connected to a "Teflon" stopcock.

4.7. Weighing bottle.

4.8. Powder funnel.

4.9. Oven regulated at 200°C.

4.10. Water bath.

4.11. Mortar.

5. REAGENTS

5.1. *n*-Hexane.

5.2. Light petroleum, b.p. 40-60°C, bromine value less than 1, free from residue.

5.3. Benzene, dry, free from thiophene and from residue.

5.4. Chloroform, pure.

5.5. Ethanol, 95 per cent (V/V), neutral to phenolphthalein.

5.6. Diethyl ether, freshly distilled, free from residue and from peroxides.

5.7. Silica gel, 0.10-0.15 mm (150-100 mesh), for chromatography, free from ether-soluble substances (*note 1*).

5.8. Sodium hydroxide, 0.05 N accurately standardized aqueous solution.

5.9. Phenolphthalein : 10 g/l solution in ethanol (5.5).

5.10. Chromic-sulphuric acid solution : e.g. dissolve 5 g of potassium dichromate in 100 ml of 40 per cent (V/V) sulphuric acid.

6. PROCEDURE

6.1. Preparation of Silica Gel

Place about 10 g of silica gel (5.7) in a tared weighing bottle (4.7) and cap immediately. Weigh to the nearest mg. Remove the cap and dry at 200°C for 2 hours into the oven (4.9). Remove from oven, cap immediately and allow to cool for 30 minutes at room temperature. Raise the cap momentarily to equalize the internal pressure with the atmosphere. Weigh again, reheat for 5 minutes at 200°C, cool and reweigh. Repeat this 5-minute drying cycle until two consecutive weights agree within 10 mg.

Calculate the per cent of water (A) in the original silica gel (5.7) by the formula :

$$A = \frac{\text{loss in mass}}{\text{mass of sample}} \times 100.$$

Calculate the amount of water (B) required to adjust the water content of the original silica gel to 5 per cent by the formula :

$$B = \frac{5 - A}{95} \times m$$

where m is the mass, in g, of the undried sample of silica gel to be used.

Weigh accurately the appropriate amount of the undried silica gel to be used in the determination, transfer to a conical flask (4.3) and add the calculated amount of water to give a final water content of 5 ± 0.1 per cent. Insert the stopper, allow to stand for at least 2 hours, while shaking occasionally to ensure complete water distribution.

Determine the water content of the adjusted silica gel as described above, and readjust, if necessary (*note 2*).

6.2. Preparation of Sample

N.B. To avoid rearrangement of partial glycerides, use extreme caution in applying heat to samples.

Melt the sample, if necessary, by warming just above the melting point for short periods, not exceeding a total of 30 minutes. Mix well.

Sample with a high melting point may be ground in a mortar (4.11) instead of being melted.

Weigh accurately about 1 g of the mixed sample into a 100-ml beaker (4.2).

Add 15 ml of chloroform (5.4) and warm, if necessary, to effect solution. Keep heating to a minimum and do not exceed 40°C (*note 3*).

6.3. Preparation of Chromatographic Column

Connect the chromatographic column (4.4) to the adapter (4.6) but do not grease the joints. Weigh 30 g of the prepared silica gel into a 150-ml beaker (4.2), add 50-60 ml of light petroleum (5.2) and stir slowly with a glass rod until all air bubbles are expelled.

Transfer the slurry to the column through a powder funnel (4.8). Open the stopcock and allow the liquid level to drop to about 2 cm above the silica gel. Transfer any silica gel slurry remaining in the beaker into the column by rinsing with a minimum amount of light petroleum. Rinse the funnel and sides of the column with light petroleum.

Drain the solvent through the stopcock until the liquid level drops to about 2 cm above the silica gel. Close the stopcock and remove the powder funnel.

6.4. Separation of the Glycerides

N.B. Keep the ambient temperature below the boiling point of the most volatile solvent used, viz. diethyl ether (34-35°C). Higher temperatures may lead to separation of the column packing and to channelling of the solvent.

Carefully add the sample solution to the prepared column. Open the stopcock and adjust the flow rate to 2 ml of effluent per minute. Rinse the sample beaker with 5 ml of chloroform (5.4) and add the rinsings to the column when the level has fallen to about 2 cm above the silica gel.

Discard the eluate.

N.B. Never allow the column to become dry on top, and maintain a flow rate of 2 ml per minute throughout the elution. If it should be necessary to interrupt the elution, do so when very little glyceride is passing through the bottom stopcock. However, avoid such interruptions, since solvents above the stopcock may cause pressure build-up, resulting in leakage through the stopcock or cracks in the silica gel packing.

Attach the separating funnel (4.5) to the column. Add 200 ml of benzene (5.3) and collect the effluent in a 500-ml flask (4.1); the eluate contains the *triglyceride fraction* (*note 4*).

When all of the benzene has been added from the separating funnel and the level in the column has dropped to about 2 cm above the silica gel, add 200 ml of a (1/9) (V/V) mixture of diethyl ether (5.6) and benzene (5.3). Collect the effluent in a second 500-ml flask (4.1); this eluate contains the *diglyceride and free fatty acid fraction* (*note 4*).

When all of the diethyl ether-benzene solvent has been added from the separating funnel and the level in the column has fallen to about 2 cm above the silica gel, add 200 ml of diethyl ether (5.6) (*note 5*). Collect the effluent in a third 500-ml flask (4.1); this eluate contains the *monoglyceride fraction* (*note 4*). In order to check the completeness of the elution, collect an additional quantity of 50 ml of diethyl ether in a fourth 500-ml flask.

Evaporate the individual fractions on a water bath (4.10) under a stream of nitrogen or dry air, or by means of a vacuum rotatory evaporator. Transfer the residues, using chloroform (5.4), quantitatively to tared 50-ml flasks (4.1) and evaporate on a water bath under a stream of nitrogen or dry air. Cool the flasks at room temperature for at least 15 minutes and weigh.

Reheat the flasks on the water bath for 5 minutes under a stream of nitrogen or dry air, cool for 15 minutes and reweigh. Repeat the 5-minute evaporation, cooling and reweighing procedure until two consecutive weights agree within 2 mg.

The weight of the fourth fraction should be less than 4 mg; if not, the elution has been incomplete and the procedure must be repeated with larger quantities of solvents.

Verification of the effectiveness of the separation by thin-layer chromatography

The purity of the three first fractions must be checked by thin-layer chromatography, using silica gel bound with plaster, ascending development with hexane (5.1)-diethyl ether (5.6) (80/20) (V/V) and visualization with chromic-sulphuric acid (5.10) at 180°C. Less than 0.5 per cent of one glyceride type in another can be detected in an application of 0.2 µl of a 10 per cent solution of the material in carbon tetrachloride.

If the separation turns out to have been unsatisfactory, repeat the procedure with different quantities of solvents (see *note 4*). Better results may also be obtained by purifying the silica gel before use (*note 1*).

Determination of the free fatty acids

Any free fatty acid, which may be present in the sample, is eluted with the diglyceride fraction. To determine the free fatty acid content of the separated diglycerides, dissolve a weighed amount of the fraction in 25 ml of warm neutral ethanol (5.5) and titrate with the sodium hydroxide solution (5.8) in the presence of phenolphthalein (5.9) (*note 6*).

7. EXPRESSION OF RESULTS

The percentages (m/m) of tri-, di- and monoglycerides are given by the formulae :

$$\text{percentage (m/m) of triglycerides} = \frac{m_1}{m} \times 100,$$

$$\text{percentage (m/m) of diglycerides} = \frac{m_2}{m} \times (100 - A)$$

$$\text{percentage (m/m) of monoglycerides} \ \frac{m_3}{m} \times 100$$

where m_1 is the mass, in g, of the fraction eluted with benzene,

m_2 is the mass, in g, of the fraction eluted with diethyl ether-benzene mixture,

m_3 is the mass, in g, of the fraction eluted with diethyl ether,

m is the mass, in g, of the test portion,

A is the percentage (m/m) of the free fatty acids (as oleic acid) in the diglyceride fraction :

$$A = \frac{V \times T \times 282}{10 \times m_4}$$

where V is the number of ml of the sodium hydroxide solution (5.8) used,

T is the exact normality of the sodium hydroxide solution (5.8) used,

m_4 is the mass, in g, of the diglyceride fraction taken for the analysis.

8. NOTES

1. Before using a batch of silica gel (5.7) it is necessary to verify that hydrocarbons are absent. This can be done by extracting 10 g of silica gel with diethyl ether, e.g. by means of a continuous extraction apparatus (Bolton-Williams type). The weight of extractable material after 4 hours of extraction should be less than 2 mg. If not, wash the silica gel several times with diethyl ether.

   If the silica gel contains iron or other soluble material, it must be treated in the following manner:

   Slurry the product with concentrated hydrochloric acid. Decant the yellowish acid solution. Repeat the treatment twice with concentrated hydrochloric acid and then several times with dilute hydrochloric acid (1.1). Wash the silica gel until neutral by slurrying several times with distilled water followed by decantation in such a way as to remove fines.

   Transfer the wet material onto a Buchner funnel fitted into a filter flask and remove the adhering water by suction. Wash the silica gel on the funnel with distilled water and finally dry at 120°C.

2. It is necessary to check every batch of adjusted silica gel to be used for its chromatographic behaviour. The best way to do this is to carry out a complete separation according to the procedure described, using a known mixture of mono-, di- and triglycerides as the sample and collecting 50 ml of eluate each time, and to check the purity of all fractions obtained by thin-layer chromatography. If the separation is not satisfactory, try other quantities of solvents and/or purify the silica gel according to *note 1*.

3. If the sample contains too much free glycerol to yield a clear solution, weigh a larger quantity and dissolve it in chloroform. Wash the solution with distilled water until free from glycerol, dry over anhydrous sodium sulphate, filter and use an aliquot for analysis.

4. To ensure the quantitative separation of all three fractions, rinse the tip of the column into the receiver with the same solvent as was used in the elution just before changing the flasks for the next eluate. The prescribed quantities of the solvents normally suffice for a good separation. If, however, one of the components of the sample is present in excessive amount it may turn out that the prescribed 200 ml of solvent, suggested to elute this fraction, is not sufficient. In that case it is advisable to use 300 ml or more so the fraction in question has been completely removed from the column.
   This can be verified by collecting some drops of the eluate on a watch glass; after evaporation of the solvent no residues should remain.

5. The addition of diethyl ether to the column often creates internal pressure, resulting in

an increased effluent flow rate and cracking in the silica gel packing before the monoglyceride fraction has been completely eluted. Avoid this by loosening the 250-ml separating funnel from the column and letting the solvent flow into the column at the same rate as the effluent is collected.

6. It may be found that the free fatty acids present are not eluted completely with the diglyceride fraction but that part of them is recovered in the monoglyceride fraction. In that case determine the free fatty acid content of the separated monoglycerides in the same way as described for the diglyceride fraction. To check, determine the free fatty acid content of the whole sample and verify that the result matches the quantity of the free fatty acids (calculated back to the weight of the sample itself) present in the individual fractions.

Warning

Attention is drawn to the fact that benzene is a very toxic substance. It is therefore recommended that all possible precautions should be taken when carrying out the determination.

Toluene, which is less toxic, may be used instead of benzene, but in that case the procedure should be suitably modified.

2.322. Determination of Monoglycerides

(Fifth Edition : Method II.C.4)

1. SCOPE

This Standard describes a method for the determination of the amount of 1-monoglycerides in oils and fats by periodic oxidation, and expressed that as glyceryl monostearate.

2. FIELD OF APPLICATION

This Standard is applicable to animal and vegetable oils and fats.

It is applicable in the presence of free glycerol, which may be separately determined by Method III.A.1 (Fifth Edition) in the wash-waters.

3. PRINCIPLE

Oxidation of the 1-monoglycerides by an excess of periodic acid in an acetic solution. Then addition of potassium iodide and determination of the excess of periodic acid by a standardized sodium thiosulphate solution.

4. APPARATUS

4.1. 500-ml flasks, with ground-glass stoppers.

4.2. 2-l flasks, with ground-glass stoppers.

4.3. 100-ml volumetric flasks.

4.4. 50-ml volumetric pipette.

5. REAGENTS

5.1. Chloroform, analytical reagent quality.

5.2. Glacial acetic acid, analytical reagent quality.

5.3. Periodic acid, acetic solution : weigh out 5.4 g of periodic acid into a flask (4.2) and dissolve in 100 ml of distilled water. Add to this aqueous solution 1900 ml of acetic acid (5.2). Mix thoroughly and keep away from light.

5.4. Citric acid, 20 g/l aqueous solution.

5.5. Sodium thiosulphate, 0.1 N aqueous solution accurately standardized.

5.6. Starch solution, 10 g/l aqueous dispersion, recently prepared from natural starch.

6. PREPARATION OF THE SAMPLE

The sample must be homogeneous and to ensure homogeneity the fat should be melted and vigorously shaken. If the sample is solid but apparently homogeneous it should be liquified at a temperature not exceeding 10°C above its melting point.

Do not take the portion for analysis until the whole of the sample has been liquefied and well shaken. The amount to be taken for analysis can be determined in the light of its probable 1-monoglyceride content as shown in the following table :

Presumed content of 1-monoglycerides present in the sample, per cent (m/m)	Weight of the portion to be taken for assay (g)
100	0.3
75	0.4
50	0.6
40	0.7
30	1.0
20	1.5
10	3.0
5	6.0
3 or less	10.0

7. PROCEDURE

Weigh out the quantity for analysis with an accuracy of 0.1 per cent, and transfer it quantitatively into a volumetric flask (4.3) using successive small quantities of chloroform (5.1) to aid the transfer until the volume reaches about 50 ml. Add 25 ml of aqueous citric acid solution (5.4) in order to decompose any soap that may be present, shake vigorously for 1 minute and allow it to stand.

After the mixture has separated into two phases remove the supernatant aqueous phase as completely as possible and transfer it into another volumetric flask (4.3). (The washing of the chloroform solution may be effected by a suitable contrivance which allows the washing to be carried out with ease and without any risk of losing any of the aqueous or chloroform phases.)

Repeat the washing of the chloroform phase twice more, either with 25 ml of water or with 25 ml of the aqueous citric acid solution when emulsions are formed.

Combine the aqueous solutions in the second volumetric flask and fill up to the 100-ml mark with distilled water. Determine the free glycerol in this solution by taking an aliquot and proceeding as in Method III.A.1. (Fifth Edition).

Make up the chloroform solution remaining in the first volumetric flask to the 100-ml mark with chloroform.

Pipette exactly 50 ml of periodic acid solution (5.3) and 50 ml of the chloroform solution into a glass-stoppered flask (4.1). Shake for a few minutes, stopper the flask and allow to stand away from light at room temperature for 30 minutes.

Carry out a blank test under the same conditions using 50 ml of the periodic acid solution and 50 ml of chloroform.

After the 30-minute period add about 20 ml of aqueous potassium iodide solution (5.5) both to the sample and to the blank. Allow them to stand for a further minute. Add about 100 ml of distilled water to each.

Titrate with the sodium thiosulphate solution (5.6) whilst shaking continuously in order to ensure thorough mixing. Towards the end of the titration, add about 2 ml of starch solution (5.7) as indicator and then continue titrating drop by drop with vigorous agitation until the end point is reached.

The number of ml used to titrate the sample should not be less than 80 per cent of the number of ml required for the titration of the blank. If this condition is not fulfilled, repeat the determination using a smaller quantity for analysis, to ensure that the oxidation of the 1-monoglycerides is complete.

8. EXPRESSION OF RESULTS

The percentage (m/m) of 1-monoglycerides, calculated as glyceryl monostearate (M = 358), is given by the formula :

$$\frac{35.8 \times (V_o - V_1) \times T}{m}$$

where V_o is the number of ml of the standardized sodium thiosulphate solution (5.6) utilized for the blank test,

V_1 is the number of ml of the standardized sodium thiosulphate solution (5.6) utilized for the test portion,

T is the exact normality of the standardized sodium thiosulphate solution (5.6) used,

m is the mass, in g, of the test portion.

Section II

Oils and Fats

Determination of the Other Natural Constituents

2.401. Determination of the Unsaponifiable Matter

(Fifth Edition : Method II.D.5)

1. SCOPE

This Standard describes two methods for the determination of the unsaponifiable matter in oils and fats.

2. FIELD OF APPLICATION

This Standard is applicable to animal and vegetable oils and fats. It is, however, only approximative for certain fats having a high content of unsaponifiable matter, e.g. fats of marine animals.

The percentage of unsaponifiable matter has generally to be added to the percentage moisture and impurities, therefore its determination should be made on the sample as received. In other purposes it may be necessary to express the unsaponifiable matter as a percentage of the pure fat present; this can be calculated from the figure determined on the sample as received and should be reported as a percentage on the pure fat present.

3. DEFINITION

The unsaponifiable matter is the whole of the products present in the substance analysed which, after saponification thereof with an alkaline hydroxide and extraction by a specified solvent, remains non-volatile under the defined conditions of test.

It includes lipids of natural origin such as sterols, higher aliphatic alcohols, pigments and hydrocarbons as well as any foreign organic matter non-volatile at 103°C (e.g. mineral oils) which may be present.

4. PRINCIPLE

Saponification of the fat by an ethanolic alkaline hydroxide solution, dilution, then extraction of the unsaponifiable matter by a solvent, diethyl ether or hexane.

In the case of hexane, a blank test is required, and its result is taken off from the figure of the assay.

5. DIETHYL ETHER METHOD

5.1. Foreword

The diethyl ether method is to be retained as the reference method, and should be used when it is possible to do so.

5.2. Apparatus

5.2.1. 250-ml round-bottomed flasks, with ground joints.

5.2.2. Reflux condenser to fit the flasks (5.2.1).

5.2.3. 500-ml separating funnels; teflon stopcock and teflon stopper are recommended.

5.2.4. Desiccator.

5.2.5. Boiling-water bath.

5.2.6. Oven, regulated at 103 ± 2°C.

5.2.7. Grain pumice, fat free.

5.3. Reagents

5.3.1. Ethanol, 95 per cent (V/V).

5.3.2. Diethyl ether, freshly distilled, free from peroxides and residue.

5.3.3. Acetone.

5.3.4. Potassium hydroxide, approximately 0.5 N aqueous solution.

5.3.5. Potassium hydroxide, approximately N ethanolic solution : dissolve 60 g of potassium hydroxide in 50 ml of water and make up to 1 l with ethanol (5.3.1). The solution should be colourless or straw yellow.

5.3.6. Potassium hydroxide, 0.1 N ethanolic solution, accurately standardized.

5.3.7. Phenolphthalein, 10 g/l solution in ethanol (5.3.1).

Verify that all the reagents gives no "unsaponifiable matter" under the test conditions.

5.4. Procedure

Weigh to the nearest 0.01 g about 5 g of the well-mixed sample into a flask (5.2.1).

Add 50 ml of the potassium hydroxide solution (5.3.5) and some pumice (5.2.7), attach the reflux condenser (5.2.2), and boil gently for 1 hour.

Stop heating. Add 100 ml of distilled water through the top of the condenser and swirl.

After cooling transfer the solution to a separating funnel (5.2.3). Rinse the flask and the pumice several times with diethyl ether (5.3.2) (100 ml in all) and pour this into the separating funnel. Stopper and shake vigorously for 1 minute, periodically releasing pressure by inverting the separating funnel and opening the stopcock.

Allow to stand until there is complete separation of the two phases. Then draw off the soap solution as completely as possible into a second separating funnel.

Extract the aqueous ethanolic soap solution twice more, each time in the same way with 100 ml of diethyl ether. Combine the three ether extracts in one separating funnel containing 40 ml of water.

Gently rotate the separating funnel containing the combined extracts and the 40 ml water. Violent agitation at this stage may result in troublesome emulsions. Allow the layers to separate completely and draw off the lower aqueous layer. Wash the ethereal solution twice more with 40-ml portions of water, shaking vigorously each time and discarding the lower aqueous layers after separation. Draw off each washing solution up to 2 ml, then rotate the separating funnel around its axis, wait some minutes to give the last remainders the opportunity for collection and draw off the collected remainders, close stopcock when ether starts to pass the bore of the stopcock.

Wash the ethereal solution successively with 40 ml of aqueous potassium hydroxide solution (5.3.4), 40 ml of water, and again with 40 ml of aqueous potassium hydroxide solution, then at least twice more with 40 ml of water.

Continue to wash with water until the wash-water no longer gives a pink colour on the addition of a drop of phenolphthalein solution (5.3.7).

Transfer the ethereal solution quantitatively a little at a time through the top of the separating funnel into a flask (5.2.1) previously dried and weighed to the nearest 0.0001 g.

Evaporate the solvent by distillation on a boiling-water bath (5.2.5).

Add 5 ml of acetone (5.3.3) and remove the volatile solvent completely in a gentle current of air, holding the flask obliquely while turning it in a boiling-water bath.

Dry the residue in the oven (5.2.6) at 103 $\pm$ 2°C for 15 minutes, placing the flask in an almost horizontal position.

Cool in a desiccator (5.2.4) and weigh to the nearest 0.0001 g.

Repeat the drying for successive 15-minute periods until the loss of weight between two successive weighings is less than 0.0015 g.

Note : If constant mass is not obtained after three periods of drying, the unsaponifiable matter is probably contaminated.

After weighing the residue dissolve it in 4 ml of diethyl ether (5.3.2) and then add 20 ml of ethanol (5.3.1) previously neutralized to a faint pink colour, using phenolphthalein indicator

(5.3.7). Titrate with the ethanolic potassium hydroxide solution (5.3.5) to the same final colour.

Correct the weight of the residue from the free acidity content.

5.5. Expression of Results

The amount of unsaponifiable matter, in per cent (m/m), is given by the formula :

$$\frac{100\ (m_1 - 0.28VT)}{m}$$

where m is the mass, in g, of the test portion,

m_1 is the mass, in g, of the residue,

V is the number of ml of the standardized potassium hydroxide solution (5.3.6) used,

T is the exact normality of the potassium hydroxide solution (5.3.6) used.

6. HEXANE METHOD

6.1. Foreword

This hexane method differs in its procedure from the hexane method described in the Fifth Edition (Method II.D.5.5). It gives experimental figures comparable to those obtained by the preceding diethyl ether method.

6.2. Apparatus

6.2.1. 250-ml round-bottomed flasks, with ground joints.

6.2.2. Reflux condenser, to fit the flask (6.2.1).

6.2.3. 250-ml and 1000-ml separating funnels; teflon stopcocks and teflon stoppers are recommended.

6.2.4. Boiling-water bath.

6.2.5. Apparatus for drying under vacuum, e.g. rotary evaporator or similar (drying ball).

6.2.6. Grain pumice, fat-free.

6.3. Reagents

6.3.1. *n*-Hexane, free from residue, or, failing this, light petroleum b.p. 40-60°C, free from residue.

6.3.2. Ethanol, 95 per cent (V/V).

6.3.3. Ethanol, 50 per cent (V/V), exactly neutral to phenolphthalein, prepared by mixing equal volume of ethanol (6.3.2) and distilled water.

6.3.4. Acetone containing 2 per cent of water (V/V), prepared by dissolution of 2 ml of water in 98 ml of acetone.

6.3.5. Potassium hydroxide, approximately 4 N ethanolic solution, recently prepared : dissolve 60 g of potassium hydroxide in ethanol (6.3.2) and make up to 250 ml. The solution should be colourless or straw yellow.

6.3.6. Potassium hydrogencarbonate, 10 g/l aqueous solution.

6.3.7. Pyrogallol, 50 g/l ethanolic solution.

6.3.8. Hydrochloric acid, 0.01 N acetone solution, accurately standardized, prepared by dilution of 0.1 N hydrochloric acid aqueous solution with acetone.

6.3.9. Bromophenol blue, 1 g/l acetone solution : dissolve bromophenol blue in acetone, then bring to sensible tint (blue) by addition with a microburette of a 0.1 N sodium hydroxide solution.

Verify that all the reagents gives no "unsaponifiable matter" under the test conditions.

6.4. Procedure

Weigh to the nearest 0.01 g about 5 g of the well-mixed sample into a flask (6.2.1). Add 25 ml of the pyrogallol solution (6.3.7), 25 ml of the potassium hydroxide solution (6.3.5) and some pumice (6.2.6).

Attach the reflux condenser (6.2.2) and boil gently for 1 hour.

Stop heating. Add 50 ml of the potassium hydrogencarbonate solution (6.3.6) through the top of the condenser and swirl.

After cooling, transfer the lukewarm solution in a 250-ml separating funnel (6.2.3), avoiding carrying along the pumice. Rinse the flask and the pumice several times with hexane (6.3.1) (50 ml in all) and pour this into the separating funnel. Stopper and shake vigorously for 1 minute, periodically releasing pressure by inverting the separating funnel and opening the stopcock.

Allow to stand until there is complete separation of the two phases. Then draw off the aqueous ethanolic soap solution as completely as possible into a second separating funnel.

Extract the aqueous ethanolic soap layer *eight* times more, each time using the same procedure with 50 ml hexane. Decantations are very quick.

Wash the combined extracts three times with about 150 ml 50 per cent (V/V) ethanol (6.3.3) each time, shaking vigorously and drawing off the aqueous ethanolic layer after each wash. Draw off each washing solution up to 2 ml, then rotate the separating funnel around its axis, wait some minutes to give the last remainders the opportunity for collection and draw off the collected remainder; close the stopcock when the hexane layer starts to pass the bore of the stopcock.

The last washing alcohol must be colourless even after addition of an additional quantity of phenolphthalein; if this is alkaline, wash again and recheck.

Transfer the hexane solution quantitatively a little at a time through the top of the separating funnel into a flask (6.2.1) previously dried and weighed to the nearest 0.0001 g.

Evaporate the solvent by distillation on a boiling-water bath (6.2.4).

Attach the flask to the apparatus for drying (6.2.5) and dry on the boiling-water bath under the maximum vacuum of the water pump during about 15 minutes (*note 1*).

Cool to room temperature under the maximum vacuum of the water pump, carefully wipe the flask, and weigh to the nearest 0.0001 g.

Repeat the drying for successive 15-minute periods until the loss of weight between two successive weighings is less than 0.0015 g.

Note : If constant mass is not obtained after three periods of drying, the unsaponifiable matter is probably contaminated.

After weighing the residue, dissolve it in 20 ml of acetone (6.3.4) previously neutralized to a very clear yellow colour using bromophenol blue (6.3.9) indicator. Titrate with the hydrochloric acid solution (6.3.8) until same final colour. Correct the weight of the residue from soap content.

A blank test must be made, using all the same reagents.

6.5. Expression of Results

The amount of unsaponifiable matter, in per cent (m/m)., is given by the formula :

$$\frac{100\ (m_1 - 0.32VT - m_2)}{m}$$

where m is the mass, in g, of the test portion,

m_1 is the mass, in g, of the residue,

m_2 is the mass, in g, of the residue obtained with the blank,

V is the number of ml of the standardized hydrochloric acid acetone solution (6.3.8) used,

T is the exact normality of the hydrochloric acid acetone solution (6.3.8) used.

6.6. Note

1. If an apparatus for drying under vacuum is not used, dry in an oven regulated at 103 ± 2°C, placing the flask in an almost horizontal position and cooling it in a desiccator.

2.402. Qualitative Examination and Determination of the Total Sterols by Their Digitonides

(Fifth Edition : Method II.D.6)

1. SCOPE

This Standard describes a method for the characterization of the sterols and permits the determination of the origin of the fat, whether vegetable or animal, and the quantitative determination of the sterols.

2. FIELD OF APPLICATION

This Standard is applicable to animal and vegetable oils and fats.

3. DEFINITION

In this Standard the word "sterols" describes the whole of the substances which give an insoluble precipitate with digitonin under the prescribed conditions.

4. PRINCIPLE

For the qualitative examination, saponification of the fat, liberation of the fatty acids, then precipitation of the digitonides of sterols. From those, preparation and identification of the sterols acetates.

For the quantitative determination, saponification of the fat, then precipitation of the digitonides and weighing of this.

5. APPARATUS

5.1. 500-ml ground-necked round-bottomed flask.

5.2. Reflux condenser, with ground joint appropriate to the flask (5.1).

5.3. Water bath.

And for the qualitative examination :

5.4. 50-ml ground-necked conical flask.

5.5. Reflux condenser, with ground joint appropriate to the flask (5.4).

5.6. 100-ml beaker.

5.7. 50-ml separating funnel.

5.8. Buchner funnel, approx. 3 cm diameter.

5.9. Melting point apparatus.

5.10. Polarizing microscope.

And for the quantitative determination :

5.11. Sintered glass crucible, porosity G4.

5.12. Oven, regulated at 103 $\pm$ 2°C.

6. REAGENTS

6.1. Ethanol, 95 per cent (V/V).

6.2. Diethyl ether, analytical reagent quality.

6.3. Digitonin, 10 g/l solution in ethanol (6.1).

And for the qualitative examination :

6.4. Carbon tetrachloride, analytical reagent quality.

6.5. Acetic anhydride, analytical reagent quality.

6.6. Potassium hydroxide, 200 g/l solution in 70 per cent (V/V) ethanol.

6.7. Potassium hydroxide, approx. 0.5 N solution in ethanol (6.1).

6.8. Hydrochloric acid, aqueous solution (ρ = 1.13).

And for the quantitative determination :

6.9. Potassium hydroxide, 400 g/l aqueous solution.

7. QUALITATIVE EXAMINATION

7.1. Preparation of the Fatty Acids

Weigh about 50 g of the fat into a round-bottomed flask (5.1). Add 100 ml of the potassium hydroxide solution (6.6). Attach the flask to the reflux condenser (5.2) and heat over a boiling-water bath (5.3) for 15 minutes, keeping at a gentle boil and shaking occasionally. Saponification is complete soon after the liquid becomes homogeneous.

Add 150 ml of hot distilled water, then 50 ml of the hydrochloric acid solution (6.8). Leave on the water bath shaking from time to time until the separated fatty acids rise to form a clear layer.

Pour through a plain, previously moistened, filter paper. The aqueous solution filters through rapidly and the fatty acids remain on the filter.

After filtering off the aqueous solution, pierce the filter paper and pass the fatty acids through a dry folded filter paper into a beaker (5.6). They should be sufficiently dry and clear after passing through this filter.

When investigating a fat having a high melting point it may be necessary to use a hot-water jacketed funnel.

7.2. Preparation of the Digitonides

Heat the fatty acids in the beaker (5.6) to a temperature of about 80°C.

Add the digitonin solution (6.3) (about 10 ml for each 20 g of fatty acids).

Maintain the temperature at about 80°C, stirring occasionally with a glass stirring rod. A precipitate of digitonides forms.

After 30 minutes, add 20 ml of carbon tetrachloride (6.4) previously heated to about 60°C.

Filter off the crystals on a filter paper on the Buchner funnel (5.8) under vacuum.

Wash the thin cake obtained twice with 10 ml of warm carbon tetrachloride, then once with 10 ml of diethyl ether (6.2) at room temperature.

Drain the thin cake under vacuum a short while, before removing the paper from the funnel.

7.3. Preparation of the Acetates

Transfer the digitonides into a dry conical flask (5.4). Add 3 ml of acetic anhydride (6.5). Attach to the reflux condenser (5.5) and boil gently for about 15 minutes.

Disconnect the conical flask, place it over a boiling-water bath (5.3) and evaporate to dryness.

Dissolve the residue in 10 ml of hot ethanol (6.1).

Filter the solution and evaporate down to a volume of about 3 ml.

After cooling, filter the crystals formed through a filter paper in a Buchner funnel (5.8) under vacuum.

Recrystallize two or three times successively from 2 or 3 ml of ethanol (6.1). After the last crystallization collect the crystals in a small beaker and dry over a hot-water bath.

Determine their melting point with the melting point apparatus (5.9).

7.4. Preparation of the Sterols

Saponify the sterol acetates with 3 ml of the ethanolic potassium hydroxide solution (6.7).

Add about 20 ml of distilled water.

Pour into a separating funnel (5.7) and extract the sterols with about 20 ml of diethyl ether (6.2).

Wash the diethyl ether layer twice with at least 10 ml of distilled water on each occasion.

Evaporate the diethyl ether.

Dissolve the sterol residue in 1 or 2 ml of ethanol (6.1). Crystallize on a microscope slide. Note the shape of the crystals and their behaviour in polarized light under the microscope.

7.5. Evaluation of the Results

The melting point of the acetates of vegetable sterols (phytosterols) and animal sterols (cholesterol) are respectively 126-127°C and 113-115°C. Hence if the melting point of the sterol acetates obtained according to 7.3 is 117°C or higher, this indicates the presence of vegetable fat.

Phytosterol crystals are rectangular and longer than they are broad; under polarized light they show parallel extinction.

Cholesterol crystals are broader and show oblique extinction.

8. QUANTITATIVE DETERMINATION

8.1. Procedure

If "s" per cent (m/m) is the expected sterol content in the sample, weigh to within 0.1 per cent about $5/s$ g of fat into the flask (5.1).

Add $3/s$ ml of the aqueous potassium hydroxide solution (6.9), then $7/s$ ml of ethanol (6.1). Attach the reflux condenser (5.2) and bring to a gentle boil on a water bath (5.3), shaking from time to time until saponification is complete, i.e. the liquid becomes homogeneous. Continue to heat for 15 to 20 minutes.

Add 60 ml of distilled water and about 180 ml of ethanol (6.1). Adjust the temperature to about 50°C.

Add 25 to 30 ml of the digitonin solution (6.3).

Allow to cool and then stand for several hours (preferably overnight).

Filter off the crystals under vacuum on the sintered crucible (6.11) previously weighed to within 1 mg. Wash first with distilled water, then several times with ethanol (6.1). Wash finally with a few ml of diethyl ether (6.2).

Dry in an oven (5.12) at 103 ± 2°C to constant weight.

8.2. Expression of Results

The percentage content (s) (m/m) of sterols is given by the formula :

$$s = 25 \frac{a}{m}$$

where a is the mass, in g, of precipitated digitonides,

m is the mass, in g, of the test portion.

2.403. Identification and Determination of Sterols by Gas-Liquid Chromatography

(Fifth Edition : Method II.C.8)

1. SCOPE

This Standard describes a method for the identification of sterols in animal and vegetable oils and fats, and a method for the determination of these.

2. FIELD OF APPLICATION

This Standard is applicable to animal and vegetable oils and fats. It is mainly used to reveal the presence of vegetable oils and fats in animal fats. It may also be applied to detect animal fats in vegetable fats, provided account is taken of the possible original presence of small amounts of cholesterol in the latter (*note 1*). The method may, in certain circumstances, also allow the various sterols present to be determined quantitatively.

3. PRINCIPLE

Saponification of the sample; then separation of the unsaponifiable matter.

Isolation of the sterols from the unsaponifiable matter by thin-layer chromatography.

Analysis by gas-liquid chromatography of the sterol fraction so isolated (or of derivatives prepared from these) and interpretation of the chromatograms.

4. APPARATUS

4.1. Separation of the Unsaponifiable Matter

See 2.401.

4.2. Thin-layer Chromatography

4.2.1. 250-ml and 25-ml ground-necked conical flasks.

4.2.2. Reflux condenser to fit the 25-ml conical flask (4.2.1).

4.2.3. Spreader for thin-layer chromatography.

4.2.4. 20 x 20-cm glass plates for thin-layer chromatography.

4.2.5. Glass developing tank for thin-layer chromatography suitable for 20 x 20-cm plates, with ground-glass lid.

4.2.6. Spray, to spray the indicator on the plate.

4.2.7. Micropipette, or microsyringe, to deliver droplets of 0.3-0.4 ul (*note 2*).

4.2.8. Slow-filtering fluted filter paper 55 mm in diameter.

4.2.9. Desiccator.

4.2.10. Oven, maintained at 103 $\pm$ 2°C.

4.2.11. Water bath.

4.2.12. UV light, for the examination of the plates.

Prepared plates are available in commerce.

4.3. Gas-liquid Chromatography

4.3.1. Gas-liquid chromatograph, with a sensitive detector, preferably of the flame-ionization type, and a glass column.

Note : As it is known that sterols may decompose at high temperatures in the presence of metals (except silver), it is advisable to use all-glass apparatus, including the injector, the column and the insulator of the detector (quartz).

5. REAGENTS

5.1. Separation of the Unsaponifiable Matter

See 2.401.

5.2. Thin-layer Chromatography

5.2.1. Chloroform, analytical reagent quality.

5.2.2. Diethyl ether, free from peroxides and residue.

5.2.3. Silica powder, with binder, of suitable quality for thin-layer chromatography.

5.2.4. Cholesterol, 100 g/l solution in chloroform (5.2.1).

5.2.5. Rhodamine 6G, 0.5 g/l aqueous solution. Other dyes or detecting agents, such as 2',7'-dichlorofluorescein, may be used provided that they do not react with sterols.

5.3. Gas-liquid Chromatography

5.3.1. Diethyl ether, free from peroxides and residue.

5.3.2. Diatomaceous earth, particle size 150-180 μm, or 120-150 μm (80-100 mesh, or 100-120 mesh) of suitable quality for gas-liquid chromatography. Before use, the support must be freed from any traces of iron by washing with acid, and silanized.

5.3.3. Silicone rubber (SE 30, JXR, or OV 17 are especially suitable).

5.3.4. Pure sterols, e.g. cholesterol, stigmasterol, brassicasterol, β-sitosterol, campesterol, or if these are not available, the sterols recently prepared from a vegetable oil, preferably sunflower seed oil, according to 6.2 and cholesterol may be used.

6. PROCEDURE

6.1. Separation of the Unsaponifiable Matter

Prepare the sample according to 2.001, "Preparation of the sample".

Extract the unsaponifiable matter from 5 g of the prepared sample by the method described in 2.401 (diethyl ether method) but modified as follows : instead of drying the unsaponifiable matter in the oven at 103 ± 2°C, dry it at a lower temperature (not exceeding 50°C) under partial vacuum, in order to avoid any undesirable oxidative deterioration.

6.2. Isolation of the Sterols by Thin-Layer Chromatography

6.2.1. Preparation of the chromatoplates

Clean the glass plates (4.2.4) thoroughly with ethanol, light petroleum and acetone in order to remove all traces of fatty matter.

Place 30 g of silica powder (5.2.3) into a 250-ml conical flask (4.2.1). Add 60 ml of distilled water. Stopper and shake vigorously for 1 minute. Immediately pour the paste into the spreader (4.2.3). Spread in a 0.25-mm layer on the clean plates.

Dry the plates for 15 minutes in the air, and then in the oven (4.2.10) at 103 ± 2°C for 1 hour.

Cool the plates at room temperature in a desiccator (4.2.9) before use.

6.2.2. Preparation of the developing tank

Place sufficient chloroform (5.2.1) (*note 3*) into the tank (4.2.5) and fit the lid. Leave for several hours to attain the liquid/vapour equilibrium; this may be accelerated by lining three walls of the tank with filter paper with the bottom edges dipping into the solvent.

6.2.3. Isolation of the sterol fraction

Weigh the unsaponifiable matter recovered according to 6.1 and dissolve it in 10 times its weight of chloroform (5.2.1), usually 0.5 to 1.0 ml (*note 4*).

With the help of the micropipette or microsyringe (4.2.7) (*note 2*) apply 50 to 60 µl of the chloroform solution of the unsaponifiable matter in a continuous line of droplets as fine as possible 2 cm from the bottom edge of a prepared 20 x 20 glass chromatoplate according to 6.2.1, leaving a gap 2.5 cm wide from each side edge of the plate.

Apply 0.3-0.4 µl of the cholesterol solution in chloroform (5.2.4) 1 cm from the right edge and 1 cm from the left edge of the plate.

Transfer the plate immediately to the developing tank prepared according to 6.2.2, fit the lid and develop the plate until the solvent reaches 1 cm below the top edge. Remove the plate from the tank and allow the solvent to evaporate at room temperature.

Using the spray (4.2.6), spray the plate with a suitable solution, e.g. a solution of Rhodamine 6G (5.2.5) and examine the plate under ultraviolet light (4.2.12). Identify the position of the sterol fraction by comparison with the reference spot of cholesterol, and mark it with a needle.

Remove the marked band with a spatula or other convenient tool and transfer the silica so recovered into a 25-ml conical flask (4.2.1). Add 5 ml of chloroform (5.2.1), fit a reflux condenser (4.2.2) and boil gently over a water bath (4.2.11) about 15 minutes. Instead of chloroform, diethyl ether may be used : this has certain advantages, for instance Rhodamine 6G is soluble in chloroform, but insoluble in diethyl ether.

Cool, and filter into a 25-ml conical flask (4.2.1) through a fluted filter (4.2.8). Remove the silica from the filter and boil up again with 5 ml of chloroform. Repeat this treatment three times (five extractions in all).

Combine the filtrates and evaporate the solvent in a gentle current of nitrogen.

Dissolve the residue in the requisite amount (a few ml) of diethyl ether (5.2.2) (*note 5*).

This solution is then used for the analysis by gas-liquid chromatography.

6.3. Analysis of the Sterol Fraction by Gas-liquid Chromatography

6.3.1. Foreword

The method described provides a *qualitative* analysis of sterols in an oil or fat. However, it can be employed for *quantitative* examination within certain limitations :

(a) the response of each sterol is a function of the type of detector used,

(b) an accurate quantitative interpretation of the chromatograms is only possible if the individual sterols used as reference standards are pure, or in default of these, if the precise composition of the sunflower-seed oil sterols used has been determined once for all in advance by the present method, using pure sterols as standards. If this is not the case, then the results must be expressed as percentage areas of the peaks.

6.3.2. Preparation of the column

Use a glass column, or, failing this, a stainless-steel column, 180 to 200 cm long, and 2-4 mm diameter. Fill it with diatomaceous earth which has been acid washed, silanized (5.3.2) and impregnated with 2 to 5 per cent of silicone rubber (5.3.3). Condition the column carefully, for example by heating at 250°C for 48 hours, in a current of carrier gas.

It is preferable to reserve such a column for the analysis of sterols.

6.3.3. Qualitative analysis

The gas chromatographic identification of the sterols may be carried out using the sterols isolated from the unsaponifiable matter by thin-layer chromatography (as described above) or the sterols obtained according to the procedure given in Method 2.402 (*note 6*).

For the gas chromatography the following conditions must be observed :

- column temperature : 230-240°C,

- injector temperature (if this is independent of the preceding) : 20 to 30°C above that of the column,

- rate of the carrier gas : 30-50 ml N_2/min,
- resolution *R* of the peaks of campesterol and stigmasterol must be above 1.0 (*note 7*).

If this is not the case, the column must not be used for the determination of sterols.

Procedure : Prepare 20 g/l solutions of the various pure sterols (5.3.4) in diethyl ether (5.3.1) or other suitable solvent.

Inject a suitable volume of each solution into the column in order to ascertain the retention distance (or time) corresponding to each sterol.

If it is not possible to obtain the various pure sterols, and mainly when only a qualitative assay is required, the solutions of pure sterols may be replaced by a 50 g/l solution of the sterols from a vegetable oil, preferably sunflower-seed oil of known qualitative composition, plus a 20 g/l solution of cholesterol (5.3.4).

The relative retention distances (or times) of the various sterols compared with cholesterol are normally within the following ranges :

Cholesterol	1.00
Brassicasterol	1.13 - 1.15
Campesterol	1.32 - 1.34
Stigmasterol	1.44 - 1.46
β-Sitosterol	1.66 - 1.68
Δ7 Stigmastenol	1.85 - 1.95 (is heavily dependent on the nature of the stationary phase)

Then inject on to the column a suitable amount of the solution of the sterols isolated from the test sample and identify the sterols by measuring the retention distances (or times) of the peaks of the chromatogram.

6.3.4. Quantitative analysis. Expression of results

Use the method of internal standards, i.e. assume that the whole of the sterols present in the sample is represented by the chromatogram (total elution).

If pure sterols can be assayed, determine their correction factors K_i with reference to cholesterol (K = 1) in the same operating conditions as those used for the sample.

Then the content of sterol i, in per cent (m/m), is given by the formula :

$$\frac{A_i \times K_i}{\Sigma(A_i \times K_i)} \times 100$$

where A_i is the area of the peak corresponding to sterol i.

In the absence of pure sterols, or sunflower-seed oil sterols of accurately known composition, one must assume that all the correction factors K_i are equal to 1 : the results are then expressed as percentages of the areas under the peaks.

The results should be reported to the nearest unit.

7. NOTES

1. If the object is to detect the presence of animal fats in vegetable oils or fats, it is necessary to employ other analytical tests in addition to the identification of the sterols, for example, a determination of the composition of the fatty acids by gas-liquid chromatography (Method 2.302) : in practice, branched-chain acids and fatty acids with an odd number of carbon atoms are only present in significant amounts in animal fats and oils.

2. Automatic apparatus for spotting can be used.

3. If chloroform of analytical reagent quality is lacking, the thin-layer chromatographic separation may be performed with any other suitable solvent or mixture of solvents, for example a mixture of hexane/diethyl ether 3/1 (V/V), or a mixture of benzene/ethyl acetate 85/15 (V/V), or a mixture of benzene/acetone 95/5 (V/V).

If the mixture of hexane/diethyl ether is used, it is better not to line the walls of the developing tank with filter paper.

4. If chloroform of analytical reagent quality is not available, any other solvent may be used.

5. Instead of diethyl ether, ethyl acetate or carbon disulphide may be used to dissolve the sterols.

6. It is possible to perform the gas-liquid chromatography on the acetates or trimethylsilyl ethers of the sterols, since with certain columns these compounds show less tailing than the free sterols.

7. The resolution R of the peaks of campesterol and stigmasterol is given by the formula (see Method 2.302) :

$$R = \frac{2(d_{R(S)} - d_{R(C)})}{\omega_{(S)} + \omega_{(C)}}$$

where d_R is the retention distance, in mm,

ω is the width, in mm, of the peak at its base.

2.421. Determination of Phosphorus

(Fifth Edition : Method II.D.16.2)

1. SCOPE

This Standard describes a method for the colorimetric determination of phosphorus in oils and fats.

2. FIELD OF APPLICATION

This Standard is applicable to animal and vegetable oils and fats.

3. PRINCIPLE

Ignition of the test portion, then nitric attack of the ashes. Formation of a yellow phosphovanadomolybdic complex between phosphoric ions and vanadic and molybdic ions, then determination of the colour so obtained.

4. APPARATUS

4.1. 5- and 20-ml calibrated pipettes.

4.2. 40-50-ml crucible or porcelain evaporating dish.

4.3. Colorimeter or spectrophotometer suitable for observations at 460 nm.

4.4. Furnace, giving a temperature of 800 to 900°C.

5. REAGENTS

5.1. Magnesium oxide, light, pure, free from phosphorus.

5.2. Nitric acid, approximately 6 N aqueous solution.

5.3. Ammonium molybdate $[(NH_4)_6Mo_7O_{24}, 4H_2O]$, analytical reagent quality, 50 g/l aqueous solution.

5.4. Ammonium vanadate $[NH_4VO_3]$, acid aqueous solution : dissolve about 2.5 g of ammonium vanadate in about 500 ml of hot water. Cool, and then add about 20 ml of nitric acid (ρ = 1.33 g/ml). Make up to 1 l with distilled water.

5.5. Standardized solution containing 1 mg of phosphorus per ml : dissolve 1.156 (± 0.001) g of disodium hydrogen phosphate $[Na_2HPO_4, 12H_2O]$, or 0.4393 (± 0.0005) g of monopotassium hydrogen phosphate $[KH_2PO_4]$ in exactly 100 ml of distilled water.

6. PROCEDURE

6.1. Calibration

Dilute the standardized solution (5.5) in such a manner as to obtain solutions containing exactly 0.4, 0.2, 0.1, 0.05 and 0.01 mg of phosphorus per ml.

Utilize these solutions in order to plot the calibration curve of the instrument (4.3) in such a way that this curve expresses the results in mg of phosphorus per ml of the final solution.

6.2. Determination

Weigh to within 1 mg exactly 0.1 g of magnesium oxide (5.1) into the crucible or porcelain dish (4.2), and ignite. Allow to cool, then weigh, to within 1 mg, about 0.1 to 10 g of fat, according to its presumed phosphorus content.

Burn off the fat (if necessary with the assistance of a folded, ashless filter paper). Ignite to a white ash in the furnace (4.4) at 800 to 900°C.

Dissolve the magnesium-containing ash in exactly 5 ml of the aqueous nitric acid solution (5.2) with the aid of a 5-ml pipette (4.1). Add exactly 20 ml of a mixture of 10 ml of the aqueous ammonium molybdate solution (5.3) and 10 ml of the acid aqueous ammonium vanadate solution (5.4). Mix and allow to stand for 20 minutes.

Prepare a blank test, not containing fat, under exactly the same conditions.

Transfer the test solution into the cell of the apparatus (4.3). Measure the extinction at 460 nm against the blank solution.

Read the absorbance (or other indications on the scale of the colorimeter).

7. EXPRESSION OF RESULTS

With the aid of the calibration curve obtained according to 6.1 and the absorbance or other figure read on the instrument, read from the graph the amount of phosphorus m_1 in mg per ml of test solution.

The percentage (m/m) of phosphorus present in the fat is given by the formula :

$$\frac{2.5\ m_1}{m}$$

where m_1 is the concentration of phosphorus, in mg per ml, read from the curve,

m is the mass, in g, of the test portion.

2.422. Determination of the Phosphoric Acid Test (PAT) Value of Raw Linseed Oil

(Fifth Edition : Method II.C.6)

1. SCOPE

This Standard describes a method for the estimation of the quantity of phospholipids in raw linseed oil.

2. FIELD OF APPLICATION

This Standard is applicable to raw linseed oil.

3. DEFINITION

The phosphoric acid test (PAT) value is the percentage of the products separated from raw linseed oil by the action of phosphoric acid according to the described method.

4. PRINCIPLE

Treatment of the test portion with 85 per cent phosphoric acid. Separation by centrifuging of the precipitated material. Washing free of oil with acetone. Drying and weighing.

5. APPARATUS

5.1. 100-ml centrifuge tubes, pear-shaped, conforming to dimensions given in the diagram 1, and made of thoroughly annealed glass. The tube is fitted with a stopper.

5.2. 1-ml pipette, graduated in 0.1 ml.

5.3. 30-ml sintered-glass crucibles of medium grade porosity (10 to 15 μ) (*note 1*).

Prepare the sintered glass crucibles as follows : add 0.3 to 0.6 g of filter aid to the empty crucible from the tip of a spatula. Mix into a slurry with approximately 15 ml of acetone. Remove the acetone by applying a vacuum to the filter. Dry the crucible in an oven at 103 ± 2°C to constant weight. Cool for 1 hour in a desiccator (5.6) and weigh to the nearest 0.1 mg. Store the prepared crucibles in the desiccator until they are to be used.

5.4. Agitator, consisting of a horizontal shaft suitably supported and fitted with clamps or clamping device for holding the pear-shaped centrifuge tubes (5.1). The tubes are held in such a manner that when the shaft rotates the tubes are tipped end over end, thus allowing the liquid content of the tube to mix as it flows from one end of the tube to the other. The shaft is rotated mechanically by any suitable means which allows the choice of two fixed speeds : a slow speed of 16 ± 2 rev/min and a fast speed of 32 ± 2 rev/min.

5.5. Centrifuge, which is capable of holding two or more tubes. It should have means of controlling its speed so as to give 5 N (500 gf) to 8 N (800 gf) at the tip of the tube (see Table 1, *note 2*).

5.6. Desiccator, containing a suitable desiccant such as anhydrous calcium sulphate.

5.7. Oven, regulated at 103 ± 2°C.

6. REAGENTS

6.1. Acetone, analytical reagent quality.

6.2. Phosphoric acid, 85 per cent (ρ_{20} = 1.7), analytical reagent quality.

6.3. Filter aid, of the diatomaceous silica type.

7. PROCEDURE

7.1. Preparation of Sample

Allow the sample to reach room temperature (23 ± 3°C) and then shake or mix thoroughly to ensure that all the sediment has been thoroughly dispersed. If the volatile matter content of the sample is greater than 0.25 per cent, dry by heating at 100°C under vacuum or by bubbling dry carbon dioxide or nitrogen at 100°C through the sample for 30 minutes. Cool the sample to room temperature.

7.2. Determination

Weigh into the centrifuge tube (5.1) 50 ± 0.01 g of the prepared sample and then add 0.5 ± 0.05 ml of phosphoric acid (6.2) from a pipette (5.2). Stopper the tube and tilt it so that the acid runs out of the tip and into the oil. Shake vigorously for a few seconds. Repeat the tilting and shake twice more.

Place the tube on the agitator (5.4) and mix for 5 minutes at such a speed that the whole of the acid disperses throughout the oil and the tip of the tube empties of oil at each revolution (a speed of 16 rev/min is adequate). Adjust the speed of the agitator so that intimate mixing without separation takes place (32 rev/min is satisfactory). Mix at this rate for 25 minutes.

Place the tube in the centrifuge (5.5) and spin it for 1 hour with a relative centrifugal force of 5 N (500 gf) to 8 N (800 gf) at the tip. The temperature should be maintained at approximately 25°C. This may be done by admitting air to the centrifuge casing.

Decant or siphon the supernatant oil as completely as possible into a second centrifuge tube and allow time for it to drain. Ordinarily the deposit stays in position as a compact mass when the tube is inverted. If the sediment layer is not compact (liquid), extra care must be taken to remove the oil without disturbing this layer. To this end a modified siphon can be used to advantage.

Add 25 ml of acetone (6.1) to the precipitate in the first tube and mix until any gummy material is dispersed. Use a wire to loosen such material from the tip of the tube if necessary. Then make up the volume to 100 ml with acetone and shake the tube.

Filter the acetone dispersion of precipitate obtained above through a prepared sintered-glass crucible (5.3), using a moderate vacuum and always maintaining some acetone in the crucible.

Thoroughly wash the centrifuge tube and precipitate in the crucible with four 15-ml portions of acetone, using a wash bottle. Since oil tends to creep up the sides of the crucible, care must be exercised.

Exhaust the acetone from the crucible, dry it in an oven (5.7) at 103 ± 2°C, cool it to room temperature in the desiccator (5.6) and weigh it to within 0.1 mg.

Treat the decanted or siphoned supernatant oil obtained after centrifuging above in the same way as the original oil, commencing from "add 0.5 ± 0.05 ml of phosphoric acid...". Reject the oil decanted or siphoned from this test.

8. EXPRESSION OF RESULTS

The phosphorus acid test (PAT) value, in per cent (m/m), is given by the formula :

$$PAT = 2 \times (m_1 + m_2)$$

where m_1 is the mass, in g, of sediment from 50 g of the original oil on first phosphoric acid treatment,

m_2 is the mass, in g, of sediment from supernatant oil on second phosphoric acid treatment.

N.B. Report the result to the second decimal place.

9. NOTES

1. The sintered glass filters must periodically be thoroughly cleaned. It is well to test the filtration rate of each crucible with pure acetone and then to discard any that cannot be cleaned to give satisfactory rates.

2. Calculate the speed of the rotation head as follows :

$$\text{speed (rev/min)} = 423 \times \left(\frac{\text{RCF}}{d}\right)^{\frac{1}{2}}$$

where RCF is the relative centrifugal force, in gf,

d is the diameter of swing, in cm, measured between tips of opposite tubes when in rotating position.

Table 1. Rotational speeds, applicable for centrifuges of various diameters of swing

Diameter of swing (cm)	Speed (rev/min) corresponding to RCF of 5 N (500 gf)	Speed (rev/min) corresponding to RCF of 8 N (800 gf)
30	1727	2184
32	1672	2115
34	1622	2052
36	1576	1994
38	1534	1941
40	1496	1892
42	1460	1846
44	1426	1804
46	1395	1764
48	1365	1727
50	1338	1692

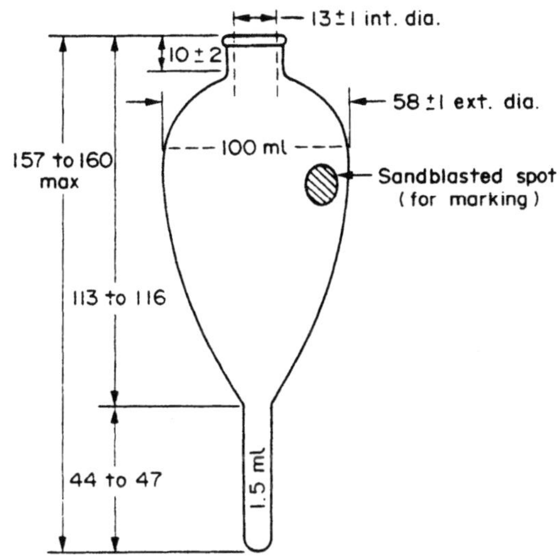

Pear-shaped centrifuge tube (dimensions are in mm).

Section II

Oils and Fats

Determination of the Quality and the Stability

2.501. Determination of the Peroxide Value (P.V.)

(Fifth Edition : Method II.D.13)

1. SCOPE

This Standard describes a method for the determination of the peroxide value of oils and fats.

2. FIELD OF APPLICATION

This Standard is applicable to animal and vegetable oils and fats.

3. DEFINITION

The peroxide value is the quantity of those substances in the sample, expressed in terms of milliequivalents of active oxygen per kg, which oxidize potassium iodide under the operating conditions described.

Note. This definition does not agree with the definition given in Method II.D.13 (Fifth Edition).

4. PRINCIPLE

Treatment of the test portion, in solution in acetic acid and chloroform, by a solution of potassium iodide. Titration of the liberated iodine with standardized sodium thiosulphate solution.

5. APPARATUS

All the equipment used shall be free from reducing or oxidizing substances.

Note. Do not grease ground surfaces.

5.1. 3-ml glass scoop.

5.2. Flasks, with ground necks and stoppers, of about 250 ml capacity, dried beforehand and filled with a pure, dry inert gas (nitrogen or, preferably, carbon dioxide).

5.3. 25- or 50-ml burette, graduated in 0.1 ml.

6. REAGENTS

6.1. Chloroform, analytical reagent quality, freed from oxygen by bubbling a current of pure, dry inert gas through it.

6.2. Glacial acetic acid, analytical reagent quality, freed from oxygen by bubbling a current of pure, dry gas through it.

6.3. Potassium iodide, saturated aqueous solution, recently prepared, free from iodine and iodates.

6.4. Sodium thiosulphate, 0.01 or 0.002 N accurately standardized aqueous solution, standardized just before use.

6.5. Starch solution, 10 g/l aqueous dispersion, recently prepared from natural soluble starch.

7. SAMPLE

Take care that the sample is taken and stored away from the light, kept cold and contained in completely filled glass containers, hermetically sealed with ground-glass or cork stoppers.

8. PROCEDURE

The test shall be carried out in diffuse daylight or in artificial light. Weigh in a glass scoop (5.1) or, failing this, in a flask (5.2), to the nearest 0.001 g, a mass of the sample in accordance with the following table, according to the expected peroxide value :

Expected peroxide value (m.e.q.)	Mass of test portion (g)
0 to 12	5.0 to 2.0
12 to 20	2.0 to 1.2
20 to 30	1.2 to 0.8
30 to 50	0.8 to 0.5
50 to 90	0.5 to 0.3

Unstopper a flask (5.2) and introduce the glass scoop containing the test portion. Add 10 ml of chloroform (6.1). Dissolve the test portion rapidly by stirring. Add 15 ml of acetic acid (6.2), then 1 ml of potassium iodide solution (6.3). Insert the stopper quickly, shake for 1 minute, and leave for exactly 5 minutes away from the light at a temperature from 15 to 25°C.

Add about 75 ml of distilled water. Titrate the liberated iodine with the sodium thiosulphate solution (6.4) (0.002 N solution for expected values less than 12, and 0.01 N solution for expected values above 12) shaking vigorously, using starch solution (6.5) as indicator.

Carry out two determinations on the same test sample.

Carry out simultaneously a blank test. If the result of the blank exceeds 0.05 ml of 0.01 N sodium thiosulphate solution (6.4), replace the impure reagents.

9. EXPRESSION OF RESULTS

The peroxide value (P.V.), expressed in milliequivalents of active oxygen per kg, is given by the formula :

$$P.V. = \frac{V \times T}{m} \times 1000$$

where V is the number of ml of the standardized sodium thiosulphate solution (6.4) used for the test, corrected to take into account the blank test.

T is the exact normality of the sodium thiosulphate solution (6.4) used.

m is the mass, in g, of the test portion.

Take as the result the arithmetic mean of the two determinations carried out.

2.502. Determination of the "Oxidized Acids" (Insolubilization Method)

(Fifth Edition : Method II.D.12)

1. SCOPE

This Standard describes a method for the determination of the fatty acids occurring during the autoxidation of animal and vegetable oils and fats.

2. FIELD OF APPLICATION

This Standard is applicable to animal and vegetable oils and fats, except those containing naturally oxygenated fatty acids (e.g. castor oil). It is not applicable to waxes.

3. DEFINITION

By convention "oxidized acids" are those substances which are insoluble in hexane, and soluble in ethanol under the specified conditions.

4. PRINCIPLE

Saponification of the test sample and extraction of the unsaponifiable matter according to 2.401, decomposition of the soaps by hydrochloric acid, extraction of the non-oxidized fatty acids by hexane, further of the oxidized fatty acids by ethanol. Evaporation of the solvent and weighing of the oxidized fatty acids.

5. APPARATUS

5.1. Apparatus used for the determination of the unsaponifiable matter (2.401.5.2, or 2.401.6.2).

5.2. 400-ml beaker, tall form.

5.3. 500-ml separating funnel.

5.4. Filter funnel, with a plain filter paper of 90 mm in diameter.

5.5. 500-ml porcelain dish.

5.6. Porcelain crucible, height 45 mm and diameter 40 mm.

5.7. Boiling-water bath.

5.8. Oven, regulated at 103 $\pm$ 2°C.

6. REAGENTS

6.1. Reagents used in the determination of the unsaponifiable matter (2.401.5.3 or 2.401.6.3).

6.2. *n*-Hexane or, failing this, light petroleum, b.p. 40-60°C, bromine value less than 1.

6.3. Ethanol, 95 per cent (V/V).

6.4. Diethyl ether.

6.5. Hydrochloric acid, aqueous solution about N.

7. PROCEDURE

7.1. Extraction of the Unsaponifiable Matter

Treat about 5 g of fat weighed to within 0.01 g as in the method for the determination of the unsaponifiable matter, either by diethyl ether (2.401.5) or by hexane (2.401.6) as far as the last extraction of the aqueous ethanolic soap solution with the chosen solvent and the two subsequent washings.

Pour the aqueous ethanolic soap solution and the two aqueous washings into the porcelain dish (5.5). Boil in order to remove completely the ethanol and, eventually, any dissolved diethyl ether.

7.2. Decomposition of the Soaps

Pour the contents of the dish quantitatively into the separating funnel (5.3), rinsing the dish with small amounts of distilled water, to make a total volume in the funnel of about 150 ml. After cooling, add a slight excess of hydrochloric acid (6.5). Shake for 2 minutes; verify that the solution is acid, adding a little more hydrochloric acid if necessary.

7.3. Separation of the Oxidized Acids by Hexane

Add 100 ml of hexane (6.2) into the funnel. Shake for a minute, opening the stopcock 2 or 3 times in order to lower the pressure.

Allow to stand for 12 hours.

Run off the aqueous acid layer. Filter the hexane solution through the filter funnel with filter paper (5.4).

The oxidized acids generally adhere to the sides of the separating funnel in a reddish brown mass. If there is a large quantity, pour the hexane from the top of the separating funnel in order to avoid blocking the stopcock. Rinse the separating funnel twice using 25 ml of hexane each time and passing each wash through the filter paper.

Then wash with hexane as follows:

- the delivery tube of the separating funnel and the filter funnel with 10 ml,
- the filter with 10 ml,
- the stem of the filter funnel with 5 ml.

Wipe the outside of the filter funnel stem and the delivery tube of the separating funnel.

7.4. Dissolution of the Oxidized Acids in Ethanol

Dissolve the oxidized acids contained in the separating funnel in hot ethanol (6.3), first using 25 ml and then 50 ml. Pass each hot solution through the filter and collect in the beaker (5.2).

Then wash the delivery tube of the separating funnel, the filter and the stem of the funnel, using 5 ml of hot ethanol for each washing, and collect in the beaker.

7.5. Evaporation of the Ethanol and Weighing of the Oxidized Acids

Evaporate the ethanolic solution until the solution has been concentrated to a small volume (a few ml).

Transfer the residue quantitatively into the tared crucible (5.6) with the aid of small amounts of diethyl ether (4.6) (this avoids the difficulties presented by the use of ethanol, which has a tendency to creep up the sides of the crucible when evaporated).

Evaporate the solvent, first in a current of air, and then over a boiling-water bath (5.7) until the ether and ethanol smells have completely disappeared, and an acrid smell is detected.

Place the crucible in the oven (5.8) at 103 $\pm$ 2°C for 30 minutes.

Weigh. Replace the crucible in the oven for a further 30 minutes, and continue to heat for 30-minute periods until the weight is constant (1 mg or less between two successive weighings).

Ignite the residue. Allow the crucible to cool. Weigh.

8. EXPRESSION OF RESULTS

The percentage of oxidized acids, in per cent (m/m), is given by the formula:

$$\frac{100\,(m_1 - m_2)}{m}$$

where m_1 is the mass, in g, of the crucible before ignition,

m_2 is the mass, in g, of the crucible after ignition.

m is the mass, in g, of the test portion.

2.504. Determination of the *p*-Anisidine Value (*p*-A.V.)

(Fifth Edition : Method II.D.26)

1. SCOPE

This Standard describes a method for the determination of the amount of aldehydes (principally 2-alkenals) in the oils and fats.

2. FIELD OF APPLICATION

This Standard is applicable to animal and vegetable oils and fats.

3. DEFINITION

The *p*-anisidine value is defined by convention as 100 times the optical density measured in a 1-cm cell of a solution containing 1.00 g of the oil in 100 ml of a mixture of solvent and reagent according to the method described.

4. PRINCIPLE

Reaction in an acetic solution of the aldehydic compounds in an oil and the *p*-anisidine (*note 1*), and then determination of the absorbance at 350 nm.

5. APPARATUS

5.1. 10-ml minimum test tubes, with ground-glass stoppers.

5.2. 25-ml graduated flasks.

5.3. Automatic pipette or automatic burette.

5.4. Spectrophotometer suitable for observation at 350 nm.

5.5. 1.00 (± 0.01)-cm glass cells, the two cells of each pair must be identical.

6. REAGENTS

6.1. *Iso*-octane (2,2,4-trimethylpentane), optically clear (*note 2*).

6.2. Glacial acetic acid, analytical reagent quality (*note 3*).

6.3. *p*-Anisidine, analytical reagent quality (*note 4*) 2.5 g/l solution in glacial acetic acid (6.2) (*note 5*).

7. PROCEDURE

N.B. The sample should be perfectly clear and dry (*note 3*).

Weigh 0.5 - 4.0 g of the sample, to the nearest 0.001 g, into a volumetric flask (5.2). Dissolve and dilute to volume with *iso*-octane (6.1).

Measure the absorbance (A_b) of the solution at 350 nm in a cell (5.5) with the spectrophotometer (5.4), using the reference cell filled with solvent as a blank.

Pipette exactly 5 ml of the fat solution into one test tube (5.1) and exactly 5 ml of the solvent into a second test tube. By means of an automatic pipette (5.3) add exactly 1 ml of the *p*-anisidine reagent (6.3) to each tube, and shake.

After exactly 10 minutes measure the absorbance (A_s) of the solution in the first test tube in a cell (5.5) at 350 nm, using the solution from the second test tube as a blank in the reference cell.

8. EXPRESSION OF RESULTS

The *p*-anisidine value (*p*-A.V.) is given by the formula :

$$p\text{-A.V.} = \frac{25 \times (1.2A_s - A_b)}{m}$$

where A_s is the absorbance of the fat solution after reaction with the *p*-anisidine reagent (6.3)

A_b is the absorbance of the solution of the fat,

m is the mass, in g, of the test portion.

9. NOTES

1. In the presence of acetic acid, *p*-anisidine reacts with aldehydic compounds in oils or fats. The intensity of colour of the yellowish reaction products formed depends not only on the amount of aldehydic compounds present but also on their structure. It has been found that a double bond in the carbon chain conjugated with the carbonyl double bond increases the molar absorbance four or five times. This means that 2-alkenals, especially, will contribute substantially to the value found.

2. In most cases *n*-hexane can be substituted for *iso*-octane as solvent. However, oils containing high amounts of oxidized fatty acids will not dissolve completely in *n*-hexane. For such oils *iso*-octane should be used as the solvent.

The absorbance of the solvent used (*iso*-octane or *n*-hexane), measured in a 1.00-cm cell between 300 and 380 nm, must be nil or nearly nil. The commercial product can be freed from absorptive material by percolating it through a glass column (3-5 cm internal diameter, and 100 cm long) filled with silica gel.

3. The reaction between *p*-anisidine and aldehydes involves the formation of water. Hence, the presence of moisture in any of the reagents or in the sample leads to incomplete reaction and, consequently, low values.

Since glacial acetic acid is highly hygroscopic, it is essential to check its moisture content by a Karl Fischer determination. If the content exceeds 0.1 per cent, the acetic acid must be discarded.

4. In storage, *p*-anisidine tends to darken as a result of oxidation. A discoloured reagent can be reduced and decolorized in the following way :

Dissolve 40 g of *p*-anisidine in 1 l of water at 75°C. Add 2 g of sodium sulphite and 20 g of active carbon and stir for 5 minutes. Then filter through a double filter paper. If carbon passes through, repeat the filtration.

Cool the filtered solution to about 0°C. Allow to stand at this temperature for at least 4 hours, or, preferably, overnight. Filter off the crystallized *p*-anisidine and wash with a small amount of water at a temperature of about 0°C.

After drying in a vacuum desiccator, transfer the crystals into a brown glass bottle. If stored in the dark and at low temperature, the crystals obtained should not darken appreciably during 1 year.

5. Reagent solutions having an absorbance greater than 0.200 when measured in a 1.00-cm cell at 350 nm against *iso*-octane or *n*-hexane as a blank should be discarded.

2.505. Evidence of Purity and Deterioration from Ultraviolet Spectrophotometry

(Fifth Edition : Method II.D.23)

1. SCOPE

This Standard describes a method for the determination of the absorbance in the ultraviolet spectrum of oils and fats and affords an indication of the purity and the deterioration of this.

The autoxidation products of oils and fats, indeed, display characteristic spectra in the ultraviolet region : linoleic hydroperoxide and the conjugated dienes which may result from its decomposition show an absorption band at about 232 nm; secondary products of autoxidation and, particularly, ethylenic diketones, show an absorption band at about 268 nm.

Conjugated trienes, which may exist naturally in certain fats (e.g. premier jus tallows) or which are formed during industrial treatment (e.g. on bleaching by bleaching earths), show a triple absorption band of which the principal peak is in the neighbourhood of 268 nm, a secondary peak at about 278 nm, and minima adjacent to the principal peak at about 262 and 274 nm.

Therefore, a determination of the absorbance $E_{1cm}^{1\%}$ at 232 nm could, theoretically, afford an indication of the state of autoxidation of a fat.

Also, theoretically, the determination of the absorbance $E_{1cm}^{1\%}$ at 268 nm and of variations between 262 and 274 nm, or, perhaps, between 277 and 283 nm, would reveal the presence of secondary products of autoxidation and of conjugated trienes in fats.

2. FIELD OF APPLICATION

This Standard is applicable to animal and vegetable oils and fats.

3. PRINCIPLE

Spectrophotometric measurement of the absorbance between 220 and 320 nm of a solution of the sample in a suitable dilution in an optically transparent solvent.

4. APPARATUS

4.1. 25-ml volumetric flask (*note 1*).

4.2. Prism or grating spectrophotometer, preferably a recording instrument.

4.3. Quartz cells, 1 cm long (*note 1*).

5. REAGENTS

5.1. *Iso*-octane (2,2,4-trimethylpentane), pure, suitable for ultraviolet spectrophotometry, affording over 70 per cent transmittance at 220 nm and above 90 per cent at 240 nm when compared with distilled water, or, failing this, and for measurements at wavelengths above 230 nm, cyclohexane, pure, for ultraviolet spectrophotometry, with transmittance above 75 per cent at 230 nm.

6. PROCEDURE

Prepare the sample according to 2.001, "Preparation of the sample". Weigh accurately, to within 0.1 mg, into a volumetric flask (4.1) an amount of the prepared sample (usually about 0.2 g) such that the absorbance of the solution will be between 0.2 and 0.8.

For wavelengths at which the absorbance is outside this range, suitable solutions must be prepared so as to bring the absorbance within the limits of 0.2 and 0.8.

Dissolve the test portion in a few ml of *iso*-octane, and make up to the mark with the solvent. Mix thoroughly.

Fill a perfectly clean quartz cell (4.3) successively with the various prepared solutions and, for comparison, another identical cell with pure *iso*-octane. If a recording spectrophotometer (4.2) is used, trace the absorption curve for wavelengths between 320 and 220 nm; otherwise, measure the absorptions at wavelengths between 225 and 235 nm and between 260 and 285 nm at 2-nm intervals (and at 1-nm intervals in the regions of the peaks and minima) and plot the absorption curve.

Note the wavelengths corresponding to the absorption maxima near the wavelengths 230, 268 and 278 nm, and the absorbance at these points.

Note the wavelengths corresponding to the absorption minima around the 268 maxima (generally 262 and 274 nm) and the absorbance at these points.

In the case of a peak near 278 nm, measure the absorbance at 284 nm.

7. EXPRESSION OF RESULTS

The absorbance $E^{1\%}_{1cm}$ at the various wavelengths is given by the formula :

$$E^{1\%}_{1cm(\lambda)} = \frac{A_\lambda}{c \times d}$$

where A_λ is the absorbance measured at wavelength λ,

c is the concentration of the solution, in g per 100 ml of the solution,

d is the length of the cell, in cm.

The variation $\Delta E^{1\%}_{1cm}$ of the absorbance $E^{1\%}_{1cm}$ at the wavelength corresponding to an absorption maximum (in the neighbourhood of 268 or 278 nm) is given by :

$$\Delta E^{1\%}_{1cm\ (\lambda M)} = E^{1\%}_{1cm\ (\lambda M)} - \frac{E^{1\%}_{1cm\ (\lambda m)} + E^{1\%}_{1cm\ (\lambda m')}}{2}$$

where $E^{1\%}_{1cm\ (\lambda M)}$ is the absorbance $E^{1\%}_{1cm}$ at the wavelength λM of the peak,

$E^{1\%}_{1cm\ (\lambda m)}$ is the absorbance $E^{1\%}_{1cm}$ at the wavelength λm of one of the minimum near the peak,

$E^{1\%}_{1cm\ (\lambda m')}$ is the absorbance $E^{1\%}_{1cm}$ at the wavelength $\lambda m'$ of the other minimum near the peak.

In the case of measurements made near 230 nm, a correction must be made for absorbance due to ester groups (-0.7) or acid groups (-0.3).

8. NOTE

1. Make sure by a check test with *iso*-octane that the glassware is free from impurities showing absorptions at wavelengths between 220 and 320 nm.

2.506. Determination of Stability (the Modified "Swift Test")

(Fifth Edition: Method II.D.21)

1. SCOPE

This Standard describes a method giving an indication of the stability to oxidation of oils and fats during storage.

The test measures the time in hours required for a sample of oil or fat aerated under carefully controlled conditions (at 98.0 ± 0.2°C) to reach a peroxide value of 100 (in milli-equivalents of active oxygen per kg of fat).

This figure for the time is regarded as an index of the resistance of the fat to autoxidation.

2. FIELD OF APPLICATION

This Standard is applicable to all oils and fats of low acidity, of animal or vegetable origin, whether or not they contain antioxidants. It is not applicable to free fatty acids.

3. PRINCIPLE

Bubbling of air at a controlled flow in the fat or oil maintained at 98.0 ± 0.2°C, then determination of the peroxide value at carefully determined times. Estimation of the time, in hours, for the obtainment of a peroxide value of 100.

4. APPARATUS

4.1. Bath or heating apparatus capable of maintaining a constant temperature of 98.0 ± 0.2°C. This must be controlled by a daily check of the temperature of a sample of oil or fat in a test tube, placed successively in the various holes in the lid under the conditions of test (air-bubbling). During the determination, the temperature must be measured in a test tube containing the prescribed amount of a mineral oil which is stable under the test conditions.

4.2. Air-distributing manifold constructed of stainless steel, nickel, aluminium or glass. The capillaries must be calibrated to permit the same flow (to within ± 10 per cent) through each outlet when the total flow is adjusted to 2.33 ml per tube per second.

4.3. Air-purification train (see Diagram 1) comprising:

(A) Air inlet tube from a source of compressed air, equipped with a stainless steel needle valve.

(B) Cylindrical wash bottle, 375 mm tall and 50 mm external diameter, containing distilled water.

(C) Cylindrical wash bottle of the same size as (B), filled to a depth of about 25 cm with a solution of 2 per cent of potassium dichromate in 1 per cent sulphuric acid. The solution must be replaced after 72 hours of continuous use.

(D) Cylindrical wash bottle of the same size as (B), filled to a depth of about 25 cm with distilled water. This must be replaced with fresh distilled water as soon as any yellow tint appears.

(E) Water-cooled Allihn condenser with five bulbs, 300-mm jacket.

(F) Trap comprising a wide-mouthed 500-ml flask, filled with glass wool.

(G and H) Pressure regulating bottles, comprising bottles of the same size as (B) and containing distilled water. The pressure may also be regulated by means of a pressure-regulating valve.

(J) Air-distributing manifold (4.2).

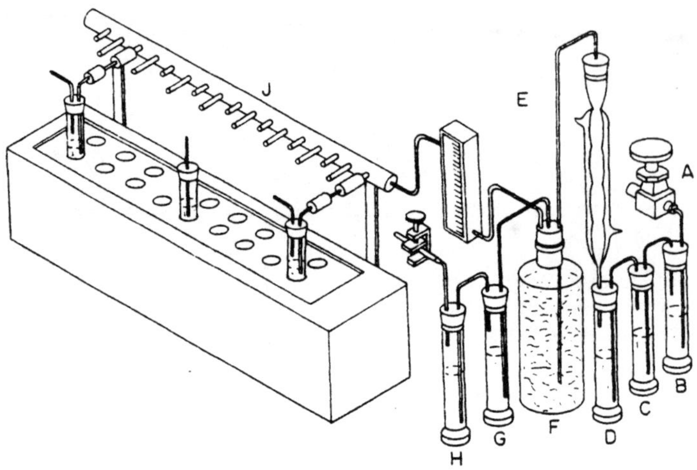

Diagram 1. Air-purification train.

4.4. Source of low pressure, clean and oil-free compressed air. An individual air compressor of the pistonless, diaphragm type is preferable to the use of industrial compressed air.

4.5. Thermometer, capable of measuring temperature to within 0.1°C between 95 and 105°C.

4.6. Gas flow meter, preferably of the calibrated cone type.

4.7. 25 x 200-mm test tubes of Pyrex glass. It is advantageous to calibrate these with a mark at the 20-ml level. Each tube is fitted with a neoprene stopper with two holes and an aeration tube (see Diagram 2). The tube used for temperature control must, additionally, be able to take the thermometer (4.5).

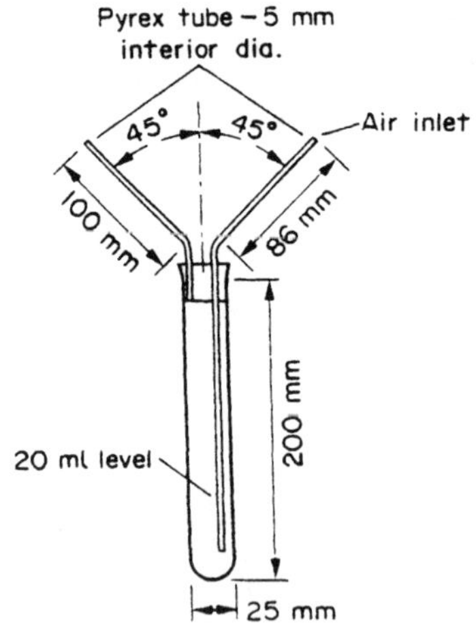

Diagram 2.

4.8. Oven, regulated at 103 ± 2°C.

4.9. Equipment for the determination of the peroxide value (see 2.501).

5. REAGENTS

5.1. Light petroleum, analytical reagent quality, b.p. 40-60°C.

5.2. Acetone, analytical reagent quality.

5.3. Detergent suitable for cleaning the apparatus, which must not leave any contaminating residues.

5.4. Reagents for the determination of the peroxide value (see 2.501).

6. PROCEDURE

6.1. Cleaning of the Test Tubes and Aeration Tubes

It is essential that the test tubes and the aeration tubes (4.7) should be most thoroughly cleaned and free from any traces of organic and inorganic matter (especially traces of metals).

After cleaning with light petroleum (5.1) and acetone (5.2), it is advisable to wash all the tubes with near boiling detergent solution (5.3), and to rinse very thoroughly.

Dry in an oven (4.8) at 103 ± 2°C, and store in a dust-free location.

6.2. Determination

Prepare the sample according to 2.001, "Preparation of the sample".

Introduce 20 ml of the prepared sample into each of two perfectly clean test tubes (4.7). Do this in such a way that none of the fat comes into contact with the upper part of the tube and consequently the stopper.

Fit the aeration tube and adjust it so that the tip of the air-delivery tube is 5 cm below the surface of the test portion.

Place the assembled tube containing the test portion in a vessel containing boiling water for 5 minutes. Remove the tube, wipe dry and place immediately in the constant-temperature heater (4.1) maintained at 98.0°C. Having previously adjusted the air-flow rate, connect the aeration tube to the capillary on the manifold (4.2) and record the starting time.

A short time before the expected end point is reached (*note 1*), determine the peroxide value according to 2.501, using a test portion of 0.3 g.

If the peroxide value so determined is above 175 m.e., the test portion must be rejected, and another determination started (*note 2*).

If the peroxide value obtained is below 75 m.e., estimate the time at which a figure of 75 m.e. would be expected, and at that time do a peroxide determination on a fresh test portion of 0.3 g. This will afford the first figure to be recorded.

Determine the peroxide value again on another test portion of 0.3 g taken from the same test tube exactly 1 hour after the first one. This affords the second value to be recorded. The two figures so obtained for the peroxide value must fall between 75 and 175 m.e.

7. EXPRESSION OF RESULTS

Taking the reaction times as abscissae, plot the two values of the peroxide value as ordinates on 2-coordinate graph paper.

Join the two points by a straight line. The stability value is then the time (in hours) at which this line cuts the ordinate for 100.

Repeat the whole operation with the second test portion and calculate the average of the results in hours (to the nearest whole number).

8. NOTES

1. After some experience, the approach of the end point can be estimated from the odour of the effluent air from the tubes.

2. It is desirable to heat and aerate the test portions continuously until the end point. When this is not practicable, remove the tube from the heating bath, chill it immediately, and store it at below 10°C, until the test tube is carried on again, starting as from 6.2, "Place the assembled tube containing the test portion in a vessel containing boiling water for 5 minutes...".

Section II
Oils and Fats
Determination of the Foreign Substances and the Additives

2.601. Determination of the Moisture and Volatile Matter

(Fifth Edition : Method II.C.1.1)

1. SCOPE

This Standard describes a method for the determination of moisture and volatile matter in oils and fats.

2. FIELD OF APPLICATION

This Standard is applicable to animal and vegetable oils and fats.

3. DEFINITION

The moisture and volatile matter content is the loss in mass which the product to be analysed undergoes on heating at 103°C, under the operating conditions specified below.

4. PRINCIPLE

Heating the test portion at 103°C on a sand-bath until moisture and volatile substances are completely eliminated.

5. APPARATUS

5.1. Dish of porcelain or glass, with flat bottom, of diameter 8 or 9 cm.

5.2. Thermometer, graduated in degrees from 80 to 120°C and of length about 10 cm.

5.3. Desiccator containing an efficient desiccant.

5.4. Sand-bath.

6. PROCEDURE

Prepare the sample according to 2.001, "Preparation of the sample".

Weigh, to the nearest 0.01 g, about 20 g of the prepared sample into the dish (5.1) which has been previously dried and then weighed with the thermometer (5.2).

Heat on a sand-bath (5.4) allowing the temperature to rise at a rate of about 10°C a minute up to 90°C and stirring continuously with the thermometer (5.2). Then reduce the rate of rise of temperature, using as a guide the release of bubbles of vapour from the bottom of the dish. Do not exceed 105°C. Continue to agitate, scraping the bottom until all evolution of bubbles has stopped.

When the amount of moisture is large, it is desirable to repeat the heating to above 103°C several times, cooling to 95°C between each heating.

Allow to cool in the desiccator (5.3) for 1 hour and then weigh.

Repeat the operations of heating, cooling and weighing until the difference between two successive weighings does not exceed 0.002 g.

Carry out two determinations on the same prepared sample.

7. EXPRESSION OF RESULTS

The moisture and volatile matter content at 103°C, as a percentage by mass, is given by the formula:

$$\frac{100\ (m - m_1)}{m}$$

where m is the mass, in g, of the test portion,

m_1 is the mass, in g, of the fat after heating.

Take as the result the arithmetic mean of the two determinations, if the conditions of repeatability are satisfied.

Report the result to one decimal place.

8. REPEATABILITY

The difference between the results of two determinations carried out simultaneously or in rapid succession by the same analyst should not exceed 0.05 g of moisture and volatile matter per 100 g of sample.

2.602. Determination of Water (Entrainment Method)

(Fifth Edition : Method II.C.1.2)

1. SCOPE

This Standard describes a method for the determination of water by azeotropic entrainment.

2. FIELD OF APPLICATION

This Standard is applicable to animal and vegetable oils and fats containing 1 per cent or more of water.

3. DEFINITION

The water content is the amount of water distilled and collected according to the method described as a percentage by mass.

4. PRINCIPLE

Determination of the amount of water carried out by distillation, with the aid of an organic liquid not miscible with water, and collected in a graduated tube.

5. APPARATUS

5.1. 500-ml short-necked flask, with ground-glass joint.

5.2. Distillation tube (see Diagram) to fit the flask (5.1) (*note 1*).

5.3. Reflux condenser, cylindrical, 400 mm effective length, to fit the tube (5.2) (*note 1*).

5.4. Boiling chips.

6. REAGENT

6.1. Commercial xylene (mixture of isomers).

7. PROCEDURE

Prepare the sample according to 2.001, "Preparation of the sample".

Weigh, to the nearest 0.01 g, 20 to 100 g of the prepared sample into the flask (5.1). Add 100 to 300 ml of xylene (6.1) and some boiling chips (5.4).

Fit the distillation tube (5.2), then the reflux condenser (5.3).

Heat gradually until the xylene boils and distils until the condensed xylene becomes clear and no more water separates from it.

Leave to stand until all water completely separates without any emulsified zone (*note 2*).

Note the volume of water.

8. EXPRESSION OF RESULTS

The water content, as a percentage by mass, is given by the formula:

$$\frac{V \times 100}{m}$$

where V is the volume, in ml, of water collected in the graduated tube,
m is the mass, in g, of the test portion.

It is assumed that the density of water is exactly 1 g/ml, within the precision of the determination.

Take as the result the arithmetic mean of two determinations if the requirements of repeatability are satisfied.

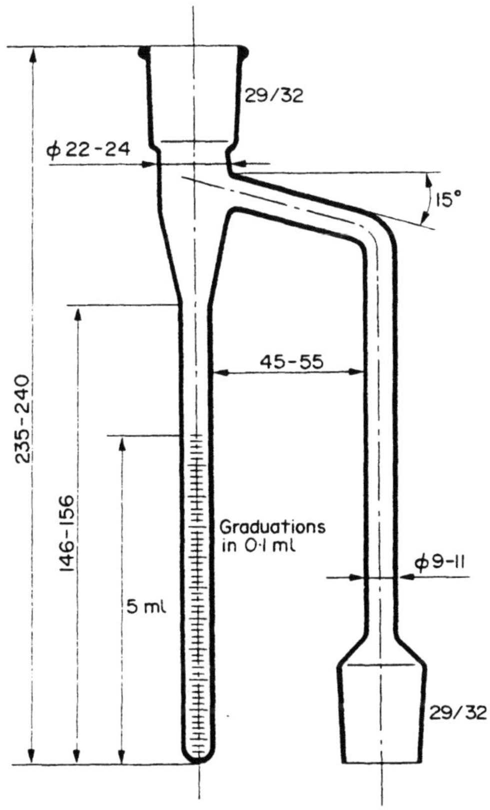

Distillation tube (dimensions are in mm).

9. REPEATABILITY

The difference between the results of duplicate determinations carried out simultaneously or in rapid succession by the same analyst, using the same mass of the test portion, should not exceed 0.1 ml of water.

The repeatability, expressed as a percentage of water, will depend on the mass of the test portion.

10. NOTES

1. First remove all trace of fat from the graduated tube and the inside of the condenser tube.

Wash successively with a chromic sulphuric acid mixture, then distilled water, and finally with acetone. Then dry the apparatus.

2. If drops of water adhere to the sides of the graduated tube, detach these by carefully heating the sides with a small flame.

2.603. Determination of Small Quantities of Water (Modified Karl Fischer Method)

(Fifth Edition : Method II.C.5)

1. SCOPE

This Standard describes a method for the determination of small quantities of water in oils and fats by a modified Karl Fischer method.

2. FIELD OF APPLICATION

This Standard is applicable to animal and vegetable oils and fats when the amount of water present is so small that the methods 2.601 and 2.602 are not sufficiently precise.

3. PRINCIPLE

Reaction between water, iodine and sulphur dioxide in the presence of an alcoholic hydroxyl group and pyridine :

$$H_2O + I_2 + SO_2 + 3C_5H_5N + ROH \rightarrow 3C_5H_6N^+ + 2I^- + RSO^-_4.$$

4. APPARATUS

4.1. 150-ml conical flasks.

4.2. 1-1 brown glass bottles, with ground-glass stoppers.

4.3. 100-ml graduated measuring cylinders, with ground-glass stoppers.

4.4. 25-ml automatic burette, protected from atmosphere moisture by a suitable dehydrant.

4.5. 25-ml graduated pipette.

4.6. Rubber stoppers, without holes, to fit the flasks (4.1).

4.7. Rubber stoppers, to fit the flasks (4.1) : pierce one hole in each stopper in such a manner that the delivery tip of the burette (4.4) can be inserted to reach to below the stopper; a desiccant is fitted in a second hole to dry any entering air.

4.8. Rubber bulb to fit the pipette (4.5).

5. REAGENTS

5.1. Methanol, anhydrous.

5.2. Chloroform, pure.

5.3. Pyridine, anhydrous.

5.4. Sulphur dioxide, anhydrous, liquid.

5.5. Iodine, pure.

5.6. Solvent mixture : mixture of chloroform (5.2) and methanol (5.1), 3/1 (V/V), stored in a bottle (4.2).

5.7. *Solution A*: dissolve 90 g of iodine (5.5) in methanol (5.1); make up to 1 l with methanol; store the solution in a bottle (4.2) at a temperature not exceeding 4°C.

5.8. *Solution B* (*note 1*): mix 450 ml of pyridine (5.3) and 450 ml of methanol (5.1) in a bottle (4.2); cool to about 4°C and, with the aid of a glass bubbling-tube, dissolve in it, a little at a time, a total of 90 g of sulphur dioxide (5.4). Protect the flask from the entry of atmospheric moisture during the operation and as far as possible maintain it at below 4°C by arresting the entry of the gas whenever the temperature shows a tendency to rise.

As soon as the addition of the sulphur dioxide is complete remove the bubbler, stopper the flask, mix well, and keep in a refrigerator at about 4°C.

5.9. *Solution C* (reagent): immediately before use pour off the desired volume of solution A into a graduated measuring cylinder (4.3) and stopper it.

Pour into another graduated measuring cylinder (4.3) an equal solution of solution B. Stopper it.

Place both cylinders in a bath of melting ice for about 15 minutes. With the aid of filter paper, dry the stoppers and upper parts of the measuring cylinders so as to eliminate all traces of water which may have condensed thereon during their stay in the ice bath. In one operation pour solution A into solution B and immediately stopper this measuring cylinder B. Shake well.

Allow to reach room temperature.

6. PROCEDURE

Throughout the determination great care must be taken to avoid the entry of atmospheric moisture.

6.1. Calibration of the Reagent (5.9)

With the aid of a rubber bulb (4.8) pipette out exactly 25 ml of the solvent mixture (5.6) into a flask (4.1). Fit a stopper (4.7) and titrate whilst stirring (a magnetic stirrer is recommended) with the solution C (5.9) until the colour changes from yellow to orange-red.

In another flask (4.1), accurately tared to within 0.0001 g, put a drop of water (*note 2*). Note, in g, to within 0.0001 g, the rise of mass. Immediately pipette exactly 25 ml of the solvent mixture (5.6). Titrate whilst shaking as described above.

The correspondence A, in g of water by ml of reagent C, is given by the formula:

$$A = \frac{m_o}{V_1 - V_o}$$

where V_o is the number of ml of the solution C used with the solvent alone,

V_1 is the number of ml of the solution C used with the solvent and the water added,

m_o is the mass, in g, of the water added.

6.2. Determination of the Water Content of Fats

Into a flask (4.1), tared to within 0.005 g, weigh accurately, to within 0.005 g, 5 to 20 g of fat. With the aid of the rubber bulk (4.8) and the pipette (4.5) pipet into the flask exactly 25 ml of the solvent mixture (5.6). Dissolve by shaking and titrate with the solution C (5.9) while shaking continuously as described in 6.1.

7. EXPRESSION OF RESULTS

The percentage (m/m) of water present in the fat is given by the formula:

$$\frac{100 \times (V - V_o) \times A}{m}$$

where V is the number of ml of the solution C used in 6.2,

V_o is the number of ml of the solution C used in 6.1 for the titration of 25 ml of solvent alone,

m is the mass, in g, of the test portion.

8. NOTES

1. The preparation of solution B is rather difficult because the sulphur dioxide gas has to be dissolved in the liquid. Operators are advised therefore to employ commercial preparations of A and B when these are separately available: these must be stored in a refrigerator at about 4°C. The mixture, solution C, should be prepared as indicated above immediately before use.

   A single reagent solution, with a somewhat different composition, is also available commercially and this may be utilized instead of the mixed reagents.

2. For the calibration of the reagent a hydrated crystalline salt (sodium tartrate, sodium acetate) may also be used instead of a drop of water.

2.604. Determination of the Insoluble Impurities

(Fifth Edition: Method II.C.2)

1. SCOPE

This Standard describes a method for the determination of insoluble impurities in raw oils and fats.

2. FIELD OF APPLICATION

This Standard is applicable to all raw animal and vegetable oils and fats.

3. DEFINITION

The insoluble impurities consist of the whole of the foreign substances which are not dissolved in the specified solvent (hexane, diethyl ether or carbon disulphide) under the conditions of the method described.

The insoluble impurities content is expressed in per cent by mass.

Impurities in edible fats and in crude fats intended for edible use shall be determined by means of hexane.

Following the experimental method described:

- hexane does not dissolve mechanical impurities (earth, sand and other debris), mineral matter, carbohydrates, nitrogenous substances, certain resins, oxidized acids, lactones, calcium soaps, and only partially dissolves hydroxylated acids and their glycerides, alkali soaps;

- diethyl ether does not dissolve mechanical impurities (earth, sand and other debris), mineral matter, carbohydrates, nitrogenous substances, certain resins, oxidized acids, lactones, calcium soaps;

- carbon disulphide does not dissolve mechanical impurities (earth, sand and other debris), mineral matter, carbohydrates, nitrogenous substances, certain resins, calcium soaps, and only partially dissolves alkali soaps.

Calcium soaps will be considered as impurities unless stipulated otherwise by agreement between the parties. In the latter case, the calcium content only, expressed as CaO, will be considered as part of the impurities.

Unless otherwise specified, only the inorganic content of alkali soaps, expressed as the metallic oxide, will be considered as part of the impurities.

4. PRINCIPLE

Treatment of the sample by an excess of solvent, filtration of the solution, washing of the filtering system with the same solvent, drying at 103 $\pm$ 2°C and weighing of the filtering system and dry residue.

5. APPARATUS

5.1. 500-ml conical flask, with ground-glass stopper.

5.2. Weighing bottle, with ground-glass stopper.

5.3. Desiccator, provided with an effective desiccant (e.g. phosphorus pentaoxide).

5.4. Oven, regulated at 103 $\pm$ 2°C.

5.5. Ashless filter paper, 12 cm in diameter.

6. REAGENTS

6.1. n-Hexane, or, failing this, light petroleum, b.p. 40-60°C, having a bromine value less than 1.

or 6.2. Diethyl ether, freshly distilled, free from peroxides.

or 6.3. Carbon disulphide, pure.

The residue on complete evaporation, for the solvent used, should not exceed 0.002 g/100 ml.

7. PROCEDURE

Prepare the sample according to 2.001, "Preparation of the sample".

Weigh accurately, to within 0.01 g, about 20 g of the prepared sample into a conical flask (5.1). Add 200 ml of the selected solvent (6.1, or 6.2, or 6.3). Insert stopper. Shake. Allow to stand at a temperature of about 20°C for 30 minutes if using hexane or diethyl ether, or for about 12 hours if using carbon disulphide.

Filter through an ashless filter paper (5.5) previously dried at 103 ± 2°C and tared into the weighing bottle (5.2) with its stopper.

Wash the filter paper with small amounts of the selected solvent, using no more than suffices to pass a fat-free filtrate (*note 1*).

Remove the filter paper from the funnel, introduce it into the weighing bottle, allow the solvent to evaporate in the open air, and complete the evaporation in the oven (5.4) at 103 ± 2°C. Remove from the oven, close the weighing bottle with its stopper, allow to cool in the desiccator (5.3), and weigh the filter with the weighing bottle and its stopper.

8. EXPRESSION OF RESULTS

The insoluble impurities content, as a percentage by mass, is given by the formula:

$$\frac{100 \times (m_2 - m_1)}{m}$$

where m is the mass, in g, of the test portion,

m_1 is the mass, in g, of the filter paper and the weighing bottle with its stopper,

m_2 is the mass, in g, of the filter paper with the insoluble impurities and the weighing bottle with its stopper.

Report the result to one decimal place.

9. NOTE

1. In certain cases, the filtrate obtained in 7 may contain ash. In this case, remove the solvent by distillation. Incinerate the residue and weigh. Add this weight to that of the impurities.

10. TEST REPORT

The test report should show the solvent used.

2.605. Determination of Ash Content

(Fifth Edition : Method II.C.3)

1. SCOPE

This Standard describes a method for the determination of the ash content of oils and fats.

2. FIELD OF APPLICATION

This Standard is applicable to animal and vegetable oils and fats.

3. DEFINITION

The ash is the mineral matter found after ignition of oils or fats.

In order to avoid the inclusion of certain elements twice, the ash content should not be added to that of the insoluble impurities as determined by method 2.604.

4. PRINCIPLE

Ignition of the test portion at a temperature lower or equal to 550°C, and weighing of the mineral residue so obtained.

5. APPARATUS

5.1. 50-ml crucible.

5.2. Ashless filter paper.

5.3. Muffle furnace.

5.4. Hot plate, or flame burner.

5.5. Boiling-water bath.

5.6. Desiccator.

6. REAGENTS

6.1. Ammonium carbonate.

6.2. Hydrogen peroxide.

7. PROCEDURE

Prepare the sample according to 2.001, "Preparation of the sample".

Weigh exactly, to within 0.01 g, about 10 g of the prepared sample into a crucible (5.1) previously tared to within 0.001 g.

Put the crucible on the hot plate (5.4), heat gently to point of ignition and allow to burn spontaneously. Ignite gently until a carbon residue is obtained which no longer emits a vapour.

Take up the residue in distilled water. Filter through an ashless filter paper (5.2). Retain the filtrate.

Place the filter paper with the residue into the crucible (5.1) and ignite in a muffle furnace (5.3) at a temperature between 500°C and 550°C until the carbon particles disappear, if necessary with the help of a gentle stream of oxygen or a few drops of hydrogen peroxide (6.2).

Allow the crucible tc cool.

Pour the retained filtrate quantitatively into it. Evaporate to dryness over a boiling-water bath (5.5). Add 0.5 to 2 g of ammonium carbonate (6.1) in order to recarbonate the ash, then ignite the residue at a temperature not exceeding 550°C.

Allow the crucible to cool in a desiccator (5.6) and weigh it to within 0.001 g.

8. EXPRESSION OF RESULTS

The ash content, in per cent (m/m), is given by the formula:

$$\frac{100 \times (m_2 - m_1)}{m}$$

where m is the mass, in g, of the test portion,

m_1 is the mass, in g, of the empty crucible (5.1),

m_2 is the mass, in g, of the crucible with the ash.

2.621. Detection of Propyl-, Octyl- and Dodecyl-Gallates, Butylated Hydroxyanisole and Butylated Hydroxytoluene

(Fifth Edition: Method II.C.9)

1. SCOPE

This Standard describes a method for the identification of some antioxidants in oils and fats.

2. FIELD OF APPLICATION

This Standard is applicable, on the one hand to animal and vegetable oils and fats, on the other hand to propyl gallate (PG), octyl gallate (OG), dodecyl gallate (DG), butylated hydroxyanisole (BHA, a mixture of 2-*tert*-butyl and 3-*tert*-butyl 4-hydroxy anisole) and butylated hydroxytoluene (BHT, 3,5-di-*tert*-butyl 4-hydroxytoluene, also called 2,6-di-*tert*-butyl *p*-cresol), all synthetic antioxidants possessing a phenolic structure (*note 1*).

3. PRINCIPLE

Dissolution of the fat in light petroleum, extraction of the antioxidants by acetonitrile, and then identification by thin-layer chromatography.

4. APPARATUS

4.1. 250-ml round-bottomed flask, which can be fitted onto the rotary evaporator (4.10).

4.2. 250- and 1000-ml conical flasks, with ground-glass stoppers.

4.3. 150-ml beaker.

4.4. 250-ml separating funnel.

4.5. 100-ml volumetric flasks.

4.6. 20-μl syringe, graduated in μl.

4.7. Glass developing tank for thin-layer chromatography, with a lid, suitable for use with 20 x 20-cm plates; the walls should be lined inside with filter paper.

4.8. Spreader for thin-layer chromatography.

4.9. Glass plates, 20 x 20 cm, for thin-layer chromatography, coated with a layer of silica gel 0.40 mm thick (*note 2*) as follows:

Thoroughly clean the glass plates with ethanol, light petroleum and acetone in order to remove all traces of fatty matter.

Put 30 g of silica powder with a binder (5.9) into a 250-ml conical flask (4.2). Add 60 ml of distilled water. Stopper and shake vigorously for 1 minute.

Pour the slurry immediately into the spreader (4.8) and spread in a layer 0.40 mm thick over the cleaned plates.

Dry the plates for 15 minutes in the air, and then at 103 ± 2°C in an oven (4.12) for an hour. Cool the plates in a desiccator (4.13) to room temperature before use.

4.10. Rotary evaporator.

4.11. Well-ventilated electric oven, regulated at 60 ± 2°C.

4.12. Well-ventilated electric oven, regulated at 103 ± 2°C.

4.13. Desiccator.

5. REAGENTS

5.1. Light petroleum, b.p. 40-60°C, free from residue, and with a bromine value less than 1.

5.2. Benzene, analytical reagent quality.

5.3. Methanol, analytical reagent quality, containing not more than 0.5 per cent (V/V) of water.

5.4. Ethanol, 95 per cent (V/V).

5.5. Chloroform, analytical reagent quality.

5.6. Glacial acetic acid, analytical reagent quality.

5.7. Ethyl acetate, analytical reagent quality.

5.8. Acetonitrile, analytical reagent quality.

5.9. Silica powder, with a binder, of suitable quality for thin-layer chromatography.

5.10. Anhydrous sodium sulphate, or anhydrous calcium chloride.

5.11. Acetonitrile saturated with light petroleum: put 900 ml of acetonitrile (5.8) into a 1000-ml conical flask (4.2). Add 100 ml of light petroleum (5.1) and a little anhydrous sodium sulphate (5.10). Shake. Stopper the flask and leave to stand overnight.

5.12. Light petroleum saturated with acetonitrile: put 900 ml of light petroleum (5.1) into a 1000-ml conical flask (4.2). Add 100 ml of acetonitrile (5.8) and a little anhydrous sodium sulphate (5.10). Shake. Stopper the flask and leave to stand overnight.

5.13. Developing solution A (*note 3*): immediately before use, prepare a mixture of light petroleum (5.1), benzene (5.2) and glacial acetic acid (5.6) in volume proportions 40/40/20.

5.14. Detector: 2,6-dichloroquinonechloroimide, 10 g/l solution in ethanol (5.4), freshly prepared: dissolve 500 mg of 2,6-dichloroquinonechloroimide in 50 ml of ethanol (5.4).

5.15. Reference solutions: 1 g/l solutions of each of the five antioxidants (PG, OG, DG, BHA, BHT) in methanol (5.3): for each antioxidant, dissolve 100 mg of the antioxidant in methanol (5.3) in a volumetric flask (4.5), make up to 100 ml with methanol and shake.

6. PROCEDURE

Prepare the sample according to 2.001, "Preparation of the sample".

6.1. Extraction of the Antioxidants

In a beaker (4.3) dissolve 7.5-10 g of the prepared sample in 100 ml of light petroleum (5.1), warming gently if necessary. Transfer the solution to a separating funnel (4.4).

Rinse the beaker with 25 ml of light petroleum, and add the rinsing to the solution in the separating funnel. No insoluble material whatever may be present in the solution or in the rinsing.

Add 25 ml of acetonitrile saturated with light petroleum (5.11) and shake for 1 minute. Withdraw the acetonitrile phase (lower layer) into a second separating funnel (4.4). If an emulsion has been formed, carefully rotate the funnel under a stream of water at 50°C until the phases become clear.

Perform similarly three more extractions, using each time 25 ml of acetonitrile saturated wtih light petroleum.

Combine the acetonitrile extracts, and wash them twice with 25-ml portions of light petroleum saturated with acetonitrile (5.12).

Transfer the washed extract to a round-bottomed flask (4.1) and fit the rotary evaporator (4.10). Evaporate the solvent under vacuum at as low temperature as possible; the temperature of the water bath must not exceed 40°C.

Dissolve the residue in 2 ml of methanol (5.3) and transfer the solution into a small storage bottle of suitable size. If the residue is not completely dissolved, filter the solution. If the amount of antioxidant is expected to be below 0.01 per cent, then use only 1 ml of methanol instead of 2 ml.

6.2. Thin-Layer Chromatography and Identification

Put sufficient developing solution (5.13) into the developing tank (4.7) to provide a layer of about 1 cm depth. Fit the lid. Leave the tank in the dark at room temperature for 1 hour or 2 in order to saturate the tank-space with the vapour of the solvent.

Activate a prepared plate (4.9) by leaving it for 1 hour in an oven (4.11) at 60 ± 2°C.

Selecting a base line 2 cm from the bottom edge of the plate, deposit 10 µl and 20 µl respectively of the methanolic solution of the extracted antioxidants in two separate spots 2 cm apart. Also, deposit 4 µl of each of the five reference solutions (5.15) in spots along the base-line 2 cm apart.

Mark a line parallel to the base-line at a distance of 15 cm. Set the plate in the developing tank and leave in the dark until the solvent front reaches the marked line. Remove the plate and allow it to dry.

Spray the plate with the detector solution (5.14) and put it in an oven (4.12) at 103 ± 2°C for 10-15 minutes.

Remove the plate from the oven and compare the R_f of the spots arising from the extract with the spots derived from the reference antioxidants (*note 4*).

Further information on the identity of the antioxidants may be obtained by putting the plate, after cooling it to room temperature, in a developing tank saturated with ammonia vapour for about 30 seconds and comparing the resulting colours (*note 4*).

7. NOTES

1. Other antioxidants having a phenolic structure, such as *nor*-dihydroguaiaretic acid (NDGA), *tert*-butyl hydroquinone (TBHQ), or the tocopherols, may also be revealed by the above method.

2. The separation can also be effected with a layer 0.25 mm thick, but sometimes it may be less sharp.

3. Supplementary information may be obtained by replacing the developing solvent A with one of the following solutions which must be made up immediately before use:

   Developing solution B: a mixture of light petroleum (5.1), benzene (5.2), ethyl acetate (5.7) and glacial acetic acid (5.6) in the proportions 40/40/25/4 by volume.

   Developing solution C: a mixture of chloroform (5.5), methanol (5.3) and glacial acetic acid (5.6) in the proportions 90/10/2 by volume.

4. As a guide, the values for R_f and the colours will be approximately as shown in the table below:

Antioxidant	Developing solution			Colour (after ammonia treatment)
	A	B	C	
PG	0.08	0.25	0.58	greyish green
OG	0.16	0.40	0.64	greyish green
DG	0.22	0.45	0.66	greyish green
BHA	0.61	0.87	0.92	violet
BHT	0.89	0.99	0.98	yellow with a violet rim

2.622. Determination of Butylated Hydroxyanisole and Butylated Hydroxytoluene

(Fifth Edition: Method II.C.10)

1. SCOPE

This Standard describes a colorimetric method for the determination of butylated hydroxyanisole (BHA, a mixture of 2-*tert*-butyl and 3-*tert*-butyl hydroxyanisole) and butylated hydroxytoluene (BHT, 3,5-di-*tert*-butyl 4-hydroxytoluene, also called 2,6-di-*tert*-butyl *p*-cresol) in oils and fats.

2. FIELD OF APPLICATION

This Standard is applicable to animal and vegetable oils and fats containing either BHA, or BHT, or a mixture of BHA and BHT, with or without other antioxidants.

3. PRINCIPLE

Separation of BHA and BHT from other antioxidants which may be present by distillation with superheated steam. In the distillate colorimetric determination of BHA by means of 2,6-dichloroquinonechloroimide, of total BHA + BHT by means of a mixture of ferric chloride and 2,2'-dipyridyl and calculation of the content of BHT by difference.

4. APPARATUS

4.1. 25-ml flasks with ground-stoppers.

4.2. 100-ml volumetric flasks.

4.3. 2-ml, 5-ml and 10-ml graduated pipettes.

4.4. All glass steam distillation apparatus as shown in Diagram, and comprising:

1. 1000-ml conical flask, with a 29/32 ground joint, containing at least 350 ml of water, and a boiling aid.
2. 13/5 spherical joint.
3. Superheating coil.
4. 1000-ml squat form beaker, containing about 500 ml of a heat stable heating fluid.
5. 13/5 spherical joint.
6. Distillation tube, with 45/40 joint.
7. 1000-ml tall form beaker, containing about 800 ml of a heat stable heating fluid.
8. 29/15 spherical joint.
9. Efficient condenser.
10. 250-ml graduated cylinder.
11. Funnel with plain filter paper.

The connections between 1 and 3 and between 3 and 6 should be lagged.

4.5. Thermometers, reading at least up to 230°C.

4.6. Spectrophotometer, or colorimeter, for measurements at 515 and 620 nm, with 1- and 2-cm cells.

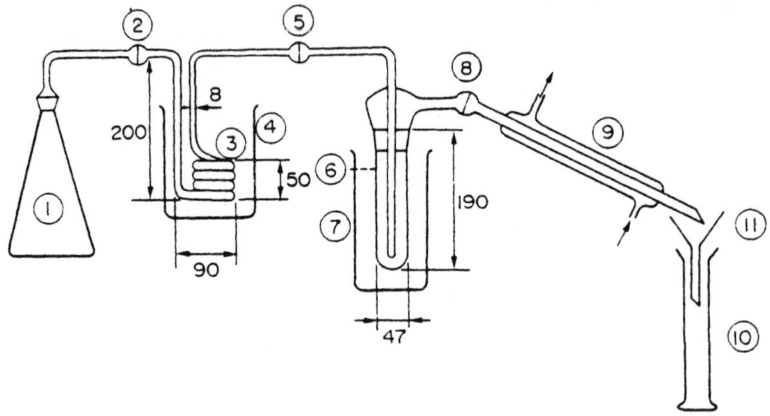

Steam distillation apparatus (dimensions in millimetres).

5. REAGENTS

5.1. Ethanol, absolute: if the blanks should show too high absorbance (e.g. over 0.15) the ethanol should be purified as follows: add 0.1 per cent of potassium hydroxide and 0.1 per cent of potassium permanganate to the absolute ethanol and distil in all-glass apparatus (no corks or rubber bungs should be used).

5.2. Ethanol, approximately 50 per cent, prepared by diluting absolute ethanol (5.1) with an equal volume of water.

5.3. *n*-Butanol, analytical reagent quality.

5.4. Butylated hydroxyanisole (BHA).

5.5. Butylated hydroxytoluene (BHT).

5.6. Anhydrous calcium chloride, analytical reagent quality.

5.7. 2,6-Dichloroquinonechloroimide, ethanolic solution: dissolve 10 mg of 2,6-dichloroquinone-chloroimide in 100 ml of absolute ethanol (5.1). The reagent must be freshly prepared each day.

5.8. 2,2'-Dipyridyl, aqueous solution: dissolve 200 mg of 2,2'-dipyridyl in about 1 ml of absolute ethanol (5.1) and dilute to 100 ml with water. Normally this reagent can be used for about a week, but if it becomes cloudy, it must be rejected.

5.9. Ferric chloride, aqueous solution: dissolve 200 mg of ferric chloride ($FeCl_3$, $6H_2O$), analytical reagent quality, in 100 ml of water. The solution must be prepared immediately before use.

5.10 Borax, aqueous solution: dissolve 2 g of borax ($Na_2B_4O_7$, $10H_2O$), analytical reagent quality, in 100 ml of water.

6. PROCEDURE

Prepare the sample according to 2.001, "Preparation of the sample".

6.1. Distillation

Put 16 g of calcium chloride (5.6) and 100 ml of water in the distillation tube 6 of the apparatus (4.4). After cooling, add 10.00 g, accurately weighed, of the prepared sample.

Before starting distillation, bring the heating bath 4 for superheated coil 3 to a temperature of 200 ± 20°C and the heating bath 7 for the distillation tube 6 to 160 ± 10°C. Maintain these two temperatures throughout the distillation.

Control the steam generator 1 so that about 5 ml per minute of water is collected in the cylinder 10 after passing through the funnel 11.

When 125 ml of distillate have been collected, stop the steam current and remove the heating bath 7 from the distillation tube 6.

After partial cooling, disconnect the condenser 9 at the joint 8, and drain off the water in the cooling jacket. Hold the condenser vertical and rinse it six times with 10-ml portions of hot absolute ethanol (5.1) (at about 60°C), passing the rinsings through the funnel 11 into the cylinder 10. Wash the filter paper with hot absolute ethanol until the total volume of the distillate and rinsings, after cooling to room temperature (*note 1*), amounts to 250 ml: this mixture is referred to as "the distillate" in the following instructions for the assay.

6.2. Determination of BHA by means of 2,6-Dichloroquinonechloroimide

All the operations must be carried out at a temperature between 20 and 25°C, and within a limited time after obtaining the distillate (a maximum of 30 minutes).

Using a graduated 10-ml pipette (4.3), transfer 10.00 ml of the distillate into a flask (4.1). Using the 2-ml graduated pipette (4.3), add 2·00 ml of the 2,6-dichloroquinonechloroimide solution (5.7) and 2.00 ml of the borax solution (5.10). Shake.

After 15 minutes add 5.00 ml of butanol (5.3). Shake. Measure the absorbance in a 1- or 2-cm cell at 620 nm, using the reagents alone in the reference cell. Let A_{620} be the value found, expressed in terms of a 1-cm cell.

Construct a reference curve, using solutions containing from 10 to 60 µg of BHA (5.4) in 10.00 ml of 50 per cent ethanol (5.2). Treat these solutions in exactly the same way as the test sample. Plot the absorbance of the various solutions against their content of BHA, and draw the graph. As the Beer-Lambert law applies to this colour reaction, the graph should be a straight line. From this straight line, derive the value of the absorbance per µg of BHA, i.e. $K_{BHA-620}$.

6.3. Determination of BHA and BHT by Means of Ferric Chloride-2,2'-Dipyridyl

Since the intensity of the coloration produce in the ferric chloride-2,2'-dipyridyl reaction varies with the intensity of the illumination and the temperature, and does not attain its maximum within the specified time of reaction, it is necessary to take special precautions. Avoid daylight as far as possible, and adhere scrupulously to the specified reaction times. The distillate and the reagents must be maintained at a constant temperature (to within ± 0.5°C) between 20 and 25°C. The reacting mixture must be maintained at the selected temperature until the absorbance is measured. The same conditions must be maintained during the analysis of the distillate and of the reference solutions.

Using a 10-ml graduated pipette (4.3), transfer 10.00 ml of the distillate into a flask (4.1). With 2-ml graduated pipettes (4.3) add 2.00 ml of the ferric chloride solution (5.9) and 2.00 ml of the 2,2'-dipyridyl solution (5.8). Shake and place the flask in the dark.

Thirty minutes after adding the ferric chloride solution, add 5.00 ml of butanol (5.3). Shake.

Thirty-five minutes after adding the ferric chloride solution, measure the absorbance at 515 nm, using the reagents alone in the reference cell. The absorbance measured represents the total of that of the BHA and BHT. Let A_{515} be the value, expressed in terms of a 1-cm cell.

Construct separate reference curves for BHA (5.4) and BHT (5.5), using solutions containing from 10 to 60 µg of one or the other compound dissolved in 10.00 ml of 50 per cent ethanol (5.2). In each case, plot the absorbance A of the various solutions against the content of BHA or BHT, and draw the two graphs. As the Beer-Lambert law applies to these two colour reactions, the graphs should be straight lines. From these derive the absorbance per µg of BHA and BHT, respectively, viz. $K_{BHA-515}$ and $K_{BHT-515}$.

7. EXPRESSION OF RESULTS

7.1. Content of BHA

The content of BHA is calculated from the assay using 2,6-dichloroquinonechloroimide.

The content α, in µg, of BHA in 10 ml of distillate, is given by the formula:

$$\alpha = \frac{A_{620}}{K_{BHA-620}}$$

and the percentage (m/m) of BHA in the oil is given by the formula:

$$\frac{\alpha \times v}{m} \times 10^{-5}$$

where m is the mass, in g, of the test portion (of the order of 10 g),

v is the number of ml of the distillate (of the order of 250 ml).

7.2. Content of BHT

The content of BHT is calculated from the results of the ferric chloride-2,2'-dipyridyl assay, taking into account the BHA content as determined above.

The content β, in μg, of BHT in 10 ml of distillate is given by the formula:

$$\beta = \frac{A_{515} - \alpha \times K_{BHA-515}}{K_{BHT-515}}$$

and the percentage content (m/m) in the oil is given by the formula:

$$\frac{\beta \times v}{m} \times 10^{-5}.$$

8. NOTE

1. On adding the ethanol, the distillate may become cloudy because traces of fatty matter have been carried over. This turbidity is eliminated later, when the butanol is added.

Made in the USA
Monee, IL
13 September 2026

62977019R10105